W9-ADT-980

APPLES
OF UNCOMMON CHARACTER

ROWAN JACOBSEN

123 HEIRLOOMS, MODERN CLASSICS, & LITTLE-KNOWN WONDERS

Photography by Clare Barboza

BLOOMSBURY

NEW YORK • LONDON • NEW DELHI • SYDNEY

Additional images by Sara Forrest (pp. vi-vii); Andrea Hubbell (pp. ii, 16); Rowan Jacobsen (p. 8); Craig Line (front endpaper; pp. 10, 12, 13)

For information address Bloomsbury USA, 1385 Broadway, New York, NY 10018.

Published by Bloomsbury USA, New York
Bloomsbury is a trademark of Bloomsbury Publishing Plc

All papers used by Bloomsbury USA are natural, recyclable products made from wood grown in well-managed forests. The manufacturing processes conform to the environmental regulations of the country of origin.

LIBRARY OF CONGRESS
CATALOGING-IN-PUBLICATION DATA

Jacobsen, Rowan.
Apples of uncommon character : 123 heirlooms, modern classics, and little-known wonders / Rowan Jacobsen; photographs by Clare Barboza.—First U.S. edition.
pages cm
Includes index.
ISBN 978-1-62040-227-6 (alk. paper)
1. Apples—Varieties. I. Barboza, Clare. II. Title.
SB363.3.A1J33 2014
634'.11—dc23

2013044004

First U.S. edition 2014

1 3 5 7 9 10 8 6 4 2

Designed and typeset by Lisa Yee

Printed and bound in China by
C&C Offset Printing Co. Ltd

Contents

INTRODUCTION

Ten years ago, my wife and I bought a neglected 1840s farmhouse in Calais, Vermont. For my wife, the attraction was the wide, worn floorboards and the classic Cape lines. For me, it was the four acres of meadows.

Both of us liked the history, the fact that you could look at any of the old maps in the town clerk's office and find our little black square with a name penciled beside it: S Laird on the 1858 Beers Atlas; TJ Porter on the 1895. Although we hadn't known any of the previous owners, the sense of continuity was particularly strong on the day in September when we closed on the house. We drove there and walked around, pinching ourselves. It was crisp and sunny, blustery with the first hints of fall, and the line of gnarled trees on the east side of the house was sporting colorful orbs of fruit. Although I hadn't noticed when we'd first looked at the house during summer, they were all apple trees.

I grew up in Vermont in the 1970s, where I'd learned that the McIntosh was the be-all and end-all of apples. Although it was a clear step up from the Red Delicious and Granny Smith apples in the supermarket, too often it was mushy and rubber-skinned, and I'd crossed apples off my life list.

But the apples hanging in these trees didn't look like any I'd ever seen in a store. In one tree, they were large, round, and striped red and yellow like little beach balls. In another, they were brown and fuzzy, more like miniature Asian pears than what I thought an apple was supposed to be. I tried one. It was strangely dry, yet very sweet, crunchy, and nutty. A third tree was full of misshapen fruits speckled with red and orange over a background swirl of greens and yellows. I picked one of these, found a relatively unscabby section, and bit into it. Juice exploded into my mouth, fragrant with cinnamon and spice. It was heavenly, and I realized right then and there that I'd been missing out.

That fall, driving the back roads of Calais, I began to notice that an extraordinary number of the trees along the roadsides were wild or abandoned apples. Every few hundred yards, the road would be scattered with little green apples, or big yellow ones, or nearly black ones. I took to sampling every tree I could reach. Quite a few were spitters, so sour and astringent that I couldn't even pretend to enjoy them, but a significant minority were not. Some tasted like pineapple, some like anise, and they were so much more interesting than apples I'd tasted before that I couldn't believe it. The world was littered with fascinating fruit! Free for the taking! It was as if an apple-centric civilization had passed from existence, and I was living amid the ruins.

Which was, in a sense, exactly what had happened. Apple culture was a huge part of eighteenth- and nineteenth-century American life. There were few national apples, but endless regional ones, each adapted to the local climate and needs—iconic apples like Rhode Island Greening and Roxbury Russet in New England; Newtown Pippin in New York and Pennsylvania; Winesap and Hewes Crab in the Southeast; Black Twig and Arkansas Black in the Mississippi River Valley; Ben Davis and Rome Beauty in the Midwest; Sierra Beauty and Gravenstein on the West Coast. Each one had been propagated because it did something superb. Some came ripe in July, some in November. Some held their shape in pies. Some started off hard and sour, but sweetened outrageously after a few months in a root cellar. Some had red flesh so full of tannins that eating one was like biting into a bar of soap, but if you pressed it and let the juice ferment in your basement all winter, it produced a dry, fragrant cider—the default buzz of agrarian America.

The New York minister and abolitionist Henry Ward Beecher, whose delightful essay on apple pie can be sampled in small servings throughout this book, described a typical mid-nineteenth-century cellar thus: "On the east side stood a row of cider barrels; for twelve or twenty barrels of cider were a fit provision for the year,—and what was not consumed for drink was expected duly to turn into vinegar, and was then exalted to certain hogsheads kept for the purpose. But along the middle of the cellar were the apple-bins; and when the season had been propitious, there were stores and heaps of Russets, Greenings, Seeknofurthers, Pearmains, Gilliflowers, Spitzenbergs, and many besides, nameless, but not virtueless."

The flavors of these apples ran the gamut, from lemon tart to pumpkin sweet, with lots of citrus, pineapple, and spice notes to bolster that classic apple scent. The shapes and colors were equally diverse. The Black Oxford looked like a plum. The Knobbed Russet looked like the love child of an apple and a toad. The Sheepnose looked like . . . well, you guessed it.

Where did such diversity come from? To our modern eyes, it all seems suspiciously like the work of some genetic engineer who took the innocent apple genome and slipped in a gene from a strawberry or a clown fish, but in fact all the genetic tinkering was done by the apple itself during its evolutionary journey from the wild forests of Kazakhstan to the fields of Europe and America. The apple has one of the largest genomes of any food plant, and it's full of recessive genes and genetic switches. In every apple seed, the genetic deck is reshuffled, new combinations of genes interact in mysterious ways, and many traits that were invisible in the parent may suddenly turn up in the child, or vice versa. What this means is that apples do not come true from seed. If you grow a tree from a McIntosh seed, it will have half the genes of its McIntosh mommy, and half the genes of its mystery dad (delivered as a pollen grain stuck to the hairy body of a bee), but its apples probably won't resemble either parent's. The apple may not fall far from the tree, but the next generation of apples will be, genetically speaking, a million miles away. Such "seedling" trees line the dirt roads, stone walls, cellar holes, and overgrown fields of rural America and rural England—and have for some time. What I was experiencing along my Vermont roadsides was strikingly similar to observations made by William of Malmesbury, a librarian in an English abbey, in the year 1125: "You may see the public highways decked with fruit trees thanks not to art and industry but to the very nature of the soil. The land produces fruit spontaneously, fruit that is far superior to all other in flavor and beauty."

Actually, most of that spontaneous fruit was fit for little more than hog feed, but every now and then, yes, a delicious new apple would come along. Since the time of the ancient Greeks and Romans, we have known that the secret to

preserving that amazing apple (or pear, plum, and most other fruit and nut trees) is by grafting: snip off a shoot and graft it onto the living rootstock of another tree. The rootstock acts simply as plumbing, pumping water and nutrients to the shoot, while the shoot, known as a "scion," carries the blueprint for all the apples produced by that tree, which will be an identical clone of the original. A few years of such grafts, and an enterprising farmer could have a whole orchard of that new apple. Then his neighbors would ask for grafts of his tasty apples, often naming them after the original farm or farmer. This is how apple varieties come into existence, and it is still how they are propagated in nurseries. Every McIntosh is a graft of a graft of the original tree that John McIntosh salvaged from the brush of his Ontario farm in 1811. Every Granny Smith stems from the chance seedling spotted by Maria Ann Smith in her Australian compost pile in 1868.

O Pioneers!

The cradle of all domestic apples is the Tian Shan, or Celestial Mountains—jagged, twenty-thousand-foot peaks that divide Kazakhstan and Kyrgyzstan from the deserts of western China. On the temperate western foothills of the Tian Shan, the apple evolved millions of years ago. It grew fat and sweet to appeal to bears and horses (still two of its biggest fans), who would eat the fruit and disperse the seeds. Then humans came along, and the apple hit pay dirt. The Silk Road that ran from imperial Rome to the Chinese capital at Xi'an cut straight through the Tian Shan passes, where wild apples easily hitchhiked in the bellies of the traders and their mounts. Soon the apple had become an integral part of Western culture.

The Romans had dozens of favorite varieties of apples, some of which—such as Court Pendu Plat and Lady—may still be with us today. The lists of Pliny the Elder, writing two thousand years ago, make it clear that the art of naming apples has changed little. The Romans had their Blood Apple, their Honey Apple, their Syrian Red and Little Greek. Monasteries carried the art of grafting through the Dark Ages, and eventually apple culture again flowered in the Renaissance. But Europe was a land of established farms, and the apple found little free room to maneuver. Favorite varieties were grafted and passed down from generation to generation. Rare was the seedling that escaped the plow and ax long enough to establish itself as a new variety. To achieve the explosion of diversity that seems to be its genetic destiny, the apple had to wait for the New World.

As they settled farms, America's colonists planted millions of apple seeds, carried with them from Europe. (In America, they found only particularly astringent crab apples, which not even the natives found terribly useful.) Most of these settlers didn't understand the basics of grafting, much less those of apple genetics, and it wouldn't have mattered if they had; apple seeds are much easier than living shoots to transport by sailing ship or wagon. Besides, most of those early trees were destined to feed the fermentation barrel or the pig, neither of which was too choosy.

Set loose in America, apples multiplied and diversified with abandon. This is what really sets the apple apart from our other food crops. Pears, plums, cherries, domestic blueberries, hybrid tomatoes, and most other fruits also do not come true from seed, but this hasn't played a big role in their history, because they don't tend to flourish when left to their own devices. Without people, they wither. Apples, on the other hand, don't need us at all. They will run rampant through any temperate environment, metamorphosing endlessly. In all our breeding work, human beings haven't teased new things out of the apple so much as pruned its potential.

While most American colonists planted seeds, a few learned men began grafting the best of the new seedling trees, and by the early 1700s nurseries selling young, grafted trees were well established, and the world's first truly American apple varieties were gaining fame. (Roxbury Russet, born in Roxbury, Massachusetts, in the early 1600s, has the distinction of being the first.) A century of seed-spitting, core-chucking, fruit sampling, and grafting took care of the rest. The Age of the Apple had arrived. In the steamy South, certain varieties showed an affinity for heat and humidity no one in Europe could ever have suspected. Thomas Jefferson planted hundreds of apple trees at Monticello, concentrating on four native varieties, two for eating (Esopus Spitzenberg and Newtown Pippin), and two for cider (Hewes Crab and Taliaferro, which has been lost), his daily drink. He had much better luck with cider than he ever had with wine.

The apple's biggest break came with the opening of the frontier, as Thoreau recognized: "Our Western emigrant is still marching steadily toward the setting sun with the seeds of the apple in his pocket, or perhaps a few young trees strapped to his load. At least a million apple-trees are thus set farther westward this year than any cultivated ones grew last year." Quite a few of those seeds were planted by John Chapman, aka Johnny Appleseed, who traveled the Ohio Valley and other Midwest regions from the 1790s through the 1830s, establishing seedling nurseries throughout the region to meet the settlers' needs. It can take ten or more years for a seedling apple tree to produce fruit, so buying a three- or four-year-old tree for six cents from Johnny Appleseed would have been a terrific way to get a head start on your homestead.

While the East Coast focused more and more on the best of the newly established varieties, propagated through grafting, the frontier continued to be a vast playground of apple seedling experimentation. John Bunker, the founder of Maine's Fedco Trees and one of the foremost champions of heirloom apples, likes to call this period the Great American Agricultural Revolution. "When this all happened," he says, "there was no USDA, no land-grant colleges, no pomological societies. This was just grassroots. Farmers being breeders." By the mid-1800s, grafting was common knowledge even on the frontier, but that still didn't stop the occasional seedling from springing into being. Nurseries such as Missouri's Stark Bro's held contests to find the best new apple varieties, and promoted the winners in their catalogs.

Farm life in the 1700s and 1800s might have been possible without apples, but apples are what made it sweet. A typical homestead might have had a dozen different apple varieties, ensuring a source of fruit and drink throughout the year, as well as vinegar for preservation. Eventually, there were more than sixteen thousand different named apples in the American nursery catalogs, though many of these were undoubtedly multiple names for the same apple. Still, the seventy-five hundred or so American varieties is several times what Europe managed to produce in three thousand years of continuous cultivation.

In the twentieth century, with the decline of the self-sufficient American farm, and the rise of industrial-scale orchards, America's apple diversity crashed. First trains, and then highways, made the national distribution of produce possible, and national distributors had little interest in working with numerous small farms having inconsistent apples. By 1920, the largest apple orchards the world had ever seen were being planted in eastern Washington State, and those apples were soon stocking the produce aisles of most supermarkets in America.

As the small family orchard became unprofitable, and Prohibition made cider apples taboo, people stopped propagating the old varieties. Of those thousands of nineteenth-century varieties, a mere six (Red Delicious, Golden Delicious, Gala, Fuji, Granny Smith, McIntosh) came to dominate the late-twentieth-century market. Not long ago, three-quarters of the apples grown in America were Red Delicious, an apple bred for its intense red color at the expense of texture and taste. Every tree was genetically identical, monocropped on a massive scale in the sun-soaked (and irrigated) deserts of eastern Washington. American consumers brought this disaster upon themselves by consistently choosing the redder apple over the tastier one.

As more and more families moved to the city, fewer had exposure to any apples except the ones in the supermarket. Children never knew that an apple could be anything more than a ball of red or green cardboard. The Age of the Apple had ended.

The Apple Sleuths

But apple trees are very patient. It's nothing for them to wait a hundred years, even two hundred. There is a bent old Black Oxford tree in Hallowell, Maine, that is 215 years old and still gives a crop of midnight-purple apples each fall.

In places like northern New England, the Appalachian Mountains, the Finger Lakes, and the Ohio Valley—agricultural hotspots that escaped the bulldozer—the old centenarians remain, on abandoned farms, or near the sign for the subdivision that went up around them, or on the orchards of a handful of preservationists who kept quietly grafting the old trees, just because. Often their identity is lost on the present landowners. "That great period of agriculture is now just far enough into the past that the last few of the old, old varieties are dying out," John Bunker says. "Are they going to be important in the future? We don't know. But they might be. And if we're going to save this stuff and give it to future generations, we'd better do it now." One of the things I hope to accomplish with this book is to familiarize people with some of these living works of art, so we can make that decision.

Bunker's love affair with apples dates to 1972, when he began farming a hardscrabble plot of land in the town of Palermo, Maine, after graduating from Colby College. That first fall, he noticed the apples ripening all over town, trees that had been planted decades ago and were now in their primes, yet mostly went ignored. So he began picking them. "I felt like these trees I was finding in my town, and then eventually all over Maine and other places, were a gift to me from someone whom I had never met, who had no idea who I was, who had no idea that I was ever going to be. Over time, I started thinking, I got to come to earth and have this amazing experience of all these trees that were grown and bearing, and all these old-timers who would take me out into their fields and show me things and take me on trips down these old roads. I would knock on somebody's door, and next thing you know I'm eating with them. It was like gift after gift after gift. And I started thinking, Do I have any responsibilities with this? Or do I just soak it up and let it go?"

So he founded Fedco Trees, which specializes in rare heirlooms, the goal being to make them less rare. When he finds a particularly delightful one, he grafts it onto rootstocks at the Fedco nursery and begins selling the trees a few years later. Over the past thirty years he has saved at least eighty varieties from oblivion. His forensic methods involve everything from studying the depth of the cavity around the stem to checking the trunk for grafting scars, to poring over old nursery catalogs and historical records. He hangs "Wanted" posters at corner stores in the towns where the apples originated, and hands them out at historical society meetings. A typical poster:

One of Bunker's best finds was the Fletcher Sweet, which he knew originated in the Lincolnville area. In 2002, he met a group from the Lincolnville Historical Society. They had never heard of the apple, but they knew of a part of Lincolnville called Fletchertown, which, like so many other old villages in northern New England, had since been reclaimed by the forest. The society posted a note in the local paper saying it was looking for an old apple called a Fletcher. A seventy-nine-year-old named Clarence Thurlow called the paper and said, "I've never heard of a Fletcher, but I know where there's a Fletcher Sweet."

Thurlow led Bunker to the dirt intersection that had once been the heart of Fletchertown, pointed to an ancient, gnarled tree, and said, "That's the tree I used to eat apples from when I was a child." The tree was almost entirely dead.

It had lost all its bark except for a two-inch-wide strip of living tissue that rose up the trunk and led to a single living branch about eighteen feet off the ground. There was no fruit. Bunker took a handful of shoots and grafted them to rootstock at his farm. A year later, both Clarence Thurlow and the tree died, but the grafts thrived, and a few years later, bore the first juicy, green Fletcher Sweet apples the world had seen in years. "It's a great apple," says Bunker. "It has a super-duper distinctive flavor." Today, Bunker has returned young Fletcher Sweet trees to Lincolnville.

This is the magic of apples. Clarence Thurlow's childhood tree was easily duplicated and returned to contemporary cultural life. Today, I can take a bite out of a Fletcher Sweet and know exactly what Clarence Thurlow was experiencing as a boy eighty years ago. I can chomp into a Newtown Pippin and understand why Thomas Jefferson exulted to James Madison from Paris that "they have no apples here to compare with our Newtown Pippin."

The New Golden Age

While John Bunker was doing his thing in Maine, others were doing the same in different regions of the country—tracking down tantalizing leads, grafting old trees, swapping specimens with their English counterparts, expanding their preservation orchards. Predictably, what they were doing had been so uncool for so long that it suddenly became very cool. We were all so bored by the sameness of supermarket fruit that the sight and taste of iconoclastic apples hit us like a jolt of tart cider.

Around the year I bought my farmhouse, my local food co-op became filled each fall with the craziest collection of apples I'd ever seen. There were tiny ones and russet ones and those unmistakable Sheepnoses, and they carried names straight out of a Dickens novel: Esopus Spitzenburg, Ashmead's Kernel, Belle de Boskoop, Ribston Pippin. The apples were being grown by a gnomic man named Zeke Goodband, who sported an earring, a waterfall of a beard like a Chinese sage's, and a lifelong devotion to apples. Zeke had transformed Scott Farm, an orchard in southern Vermont that was the home of Rudyard Kipling, into one of the great working heirloom orchards in the country. The eighty varieties at Scott Farm are, to me, as fine a collection of artistic masterpieces as you'd find in any museum, just as moving and thought-provoking, and Zeke is a superb curator. He will even let you eat his collection.

Zeke's apples are doing for market shoppers throughout New England what Fedco Trees is doing for home gardeners: Expanding our understanding of what an apple can be. That trend keeps gaining steam, and we are now in what I can confidently declare to be the Second Age of the Apple. We have more varieties of extraordinary apples within reach—through mail-order suppliers, farmers' markets, pick-your-own orchards, and enlightened grocers like Whole Foods—than any people who have come before us. Hard cider, too, is in a full-blown renaissance; there simply have never been as many superb ciders in existence, professional and amateur, as there are at this very moment.

By no means is this energy limited to heirlooms. The new generation of consumers has shown an enthusiasm for novel apples, and the commercial apple world has responded with its own outpouring of creativity. Although heirloom apple fans like to badmouth these new apples, some of the recent stars of the produce aisle, like Gala and Honeycrisp, would have been as passionately embraced by our homesteading forebears as they have been by us. They are damned fine apples, even if they have a certain generic predictability to them. All tend to be very sweet, very crisp, and very juicy. But is that so terrible?

We have come a long way from Red Delicious Land.

The first apple breeding program in the United States began at Cornell University in the 1890s, and until recently its methods were simply an accelerated version of nature's. Growers would intentionally cross-pollinate two varieties, hoping to capture the attractive qualities of both parents, but because of the apple's unpredictable genetics, to do this they had to make thousands of crosses, waiting years for the seedlings to grow large enough to bear fruit. Only then could they see what they had, propagate the best candidates, and wait many more years to test them for productiveness, disease resistance, and other commercial considerations. It took decades for a new variety to be released commercially.

Now this process goes much faster. Genetic tools enable breeders to check the DNA of a new apple seedling and see instantly which traits it has inherited from which parent, allowing them to single in on the most promising prospects very quickly. Still, this is good old-fashioned sexual reproduction (there are no genetically modified apples...yet), and it takes time: SnapDragon, Cornell's great new hope, which will hit stores in 2015, was ten years in the making, and it came into existence in the traditional way: as a seedling that proved its mettle over thousands of rivals.

In fact, as you look at the apples in this book and learn their stories, one of the things that will become clear is that each variety was a one-in-a-million shot, a double miracle: First, it had to be dealt a genetic royal flush of great attributes, and second, it had to survive long enough to produce fruit that somebody noticed. Who knows but that the next Honeycrisp is lurking not in Cornell's ordered rows but in a cellar hole in Ohio, as Thoreau recognized a century and a half ago: "Every wild-apple shrub excites our expectation thus, somewhat as every wild child. It is, perhaps, a prince in disguise." It often took a hurricane or a switched-at-birth mix-up—some unlikely act of fate that set it on its own Dickensian arc—to give each new apple variety a shot at all. Each apple prince is the hero of its own tale. Each has something different to say about being alive. And there are more to be found.

This book is here to give a tantalizing taste of the fruit with the ultimate range. Pick your way through these piles of colorful orbs, their skins aswirl in red, orange, yellow, green, and purple, and you will get a sense of the many things an apple can be, the many roles it can play in our lives. If there is one particular lesson the apple has to teach us, it is that the world is ripe with possibilities. The apple never met a landscape it couldn't partner with, never saw an opportunity it didn't relish. Like Molly Bloom, it almost always says *yes*.

THE PORTRAITS

"With an apple, I will astonish Paris!" Paul Cézanne famously boasted. He meant that he would show the world something familiar in a brand-new light, but his choice of subject was not arbitrary. "Fruits are the most loyal," Cézanne said. "They like having their portrait done." Fruits, Cézanne felt, carried with them the sum of their experience. "They come to you with their scents, speak to you of the fields they left, about the rain that nourished them, about the sunrises they observed. In capturing with fleshy strokes the skin of a beautiful peach, the melancholy of an old apple, I see that they share the same bittersweet shadow of renouncement, the same love of the sun, the same memory of dew."

No fruit conveys a more textured experience than the apple. Instead of the one-note eroticism of a plum or a peach, the typical apple seems to be struggling to be a few different things at once, and not entirely succeeding at all of them, which somehow makes it the most sympathetic of fruit. Although it has its moments of beauty, it can be pretty homely if you catch it at the wrong time. It is not without sweetness or juice, but it will never be one of the luscious ones. It tends to suffer the slings and arrows of fate right on its skin. It's a bit of a workhorse. While many other fruits breeze in and out of season, quickly embraced and quickly missed, the apple is always with you, ready to meet all your culinary needs. For this service, it is the fruit most likely to be taken for granted.

Yet all this, of course, makes the apple the most exquisite of fruit, for anyone who knows how to look. The *wabi-sabi* of its beauty etched through time, the heroic optimism of its willingness to try anything, the sheer complexity of its personality. With an apple, Cézanne may have astonished Paris, but the apple had already astonished the world.

The following portraits try to capture the apple in a few of its multifaceted moods. Every variety has its own style, and, as with a human subject, every individual will have its own changeable moods, shifting with the seasons. We haven't necessarily tried to capture a "typical" example of each variety; instead, we've gone for individual fruits that seemed to have something to say, warts and all. The descriptions try to convey the essence of each particular apple while keeping apple jargon to a minimum; if you come across an unfamiliar term or historical personage, check the glossary in the back for additional information.

I've grouped the apples into categories based on their most fitting use. This is often a judgment call, since many apples could easily have appeared in multiple categories. Keepers, for example, tend to have a firmness that makes them good pie apples in the fall. Jonagold and Northern Spy do absolutely everything well. Is McIntosh a dessert apple or a culinary apple? I listed it under culinary, because I think it is at its best in cider and sauce, but I know some who like nothing better than sinking their teeth into a fresh Mac. The main purpose of the categories is to highlight the extraordinary versatility of our most useful fruit. Apples can fold themselves into your life in a remarkable number of ways. More than that, they can bring the world into your life. The swirling colors and patterns of the seasons, the cycle of flower and fruit, nature's whole blueprint for attraction, reproduction, and immortality—it's all there in the apple, as Cézanne knew.

SUMMER APPLES

In order to have been selected for grafting by human beings and handed down from person to person through the years, an apple variety had to fill some need exceptionally well.

Having great flavor is one ticket to immortality, but so is being able to survive farther north or south than your kin, even if your flavor is only so-so. The open niche summer apples found to exploit was coming ripe a month to two months before other apples, ensuring themselves a place in the hearts of eaters who (prerefrigeration) hadn't tasted a decent apple since March.

Although summer apples come in all sorts of hues, most share certain traits: tender skin and flesh, light density, mild flavor, and a tendency to bruise. Whatever cellular chemistry it is that allows summer apples to bulk up at twice the rate of typical apples also prevents them from hanging around; the apple that burns twice as fast burns half as long. While that would be a drawback for fall or winter apples, it works just fine for the summer ones, for they are merely trying to serve as hors d'oeuvres in the apple feast.

Everything about summer apples is evanescent. The window between ripe and rot can be so narrow that you are advised to stand under the tree and eat them before they know they've been picked. The flavor, too, is light and does not linger. Thoreau praised their "fugacious ethereal qualities," though he admitted that "none of them are so good to eat as some to smell. One is worth more to scent your handkerchief with than any perfume which they sell in shops." Sometimes summer apples taste of other summer fruits, like strawberry and raspberry. Yet they are clearly "apple," and so they are forever linked with the uncomplicated joy of tasting your first apple of the year. None of the autumnal melancholy of a Snow or a Winter Sweet Paradise for these free spirits; they are children of the light and heat. Have a taste, enjoy the moment, and then move on before you tire of them, and they of you.

Chenango Strawberry

Origin Chenango Valley, New York, before 1850. **Appearance** Looks like it got stretched in Photoshop. It's shaped more like a strawberry than an apple. A hint of its five ribs can be seen in the vase shape. Its fine, translucent skin is a pale honeydew streaked with rose on the sunny side, becoming red later in the season. **Flavor** Mildly sweet and tart, Chenango Strawberry when fully ripe emits a delicious strawberry-candy perfume you can smell across the room. **Texture** Tender to the bone. Straight off the tree, its pure white flesh practically melts away. This isn't to everyone's taste, but think of it like a ripe, apple-flavored pear and you may come around. The delicate skin, with its seaweedlike pop, is a joy to eat. **Season** August to early September. Comes ripe in waves over a series of weeks. **Use** Eat right away or make elegant sauce. **Region** Home orchards in the Northeast and South.

The poster child for commercially hopeless apples, Chenango Strawberry makes a strong case for the value of backyard orchards and farmstands, because that is the only way to enjoy one. "This is an apple you'll never see on a supermarket shelf," says Brad Koehler of Windfall Orchards in Cornwall, Vermont. I picked Chenango Strawberries with Brad one sunny day in late August. "I can barely get them off my tree and to the farmers' market three miles from here," he says. "Many are so soft you can put your thumb right through them." And Brad's litany of grievances doesn't stop there. "It's such a pain. They're aggressive growers, which means a lot of pruning, and they horribly overbear," which means daily picking to get the ripe fruit off the limbs. "This apple is an act of love. There's no other reason for it to exist."

Yet exist it does, which is a testament to that love. The Chenango is the quirky flower that rewards the gentle hand, the stay-at-home beauty that doesn't travel. Savor its fragility, and give thanks for the shy things of the world.

Ginger Gold

Origin Nelson County, Virginia, 1969. **Appearance** A smooth, consistent pale green or yellow, depending on where it's grown and how ripe it was picked. It is strongly ribbed and conic, with uniform green dots over the entire surface. **Flavor** In the South and West, it tends to be sweet and simple, while in the North it impersonates Granny Smith. **Texture** Crisp, but not crackly, and very juicy. **Season** July (California), August (South), September (Northeast). **Use** Decent fresh, in pies and crisps, or in salads, where it won't brown. **Region** The best specimens are grown down south, but Ginger Gold has been embraced by orchards everywhere and is ubiquitous in summer and early fall.

In 1969, Hurricane Camille blasted up the state of Mississippi as one of the most powerful hurricanes ever recorded, then hung a right turn and blasted out through the state of Virginia, causing unprecedented flooding in the Blue Ridge Mountains, including to the orchard of Clyde and Ginger Harvey. Clyde surveyed the rutted ruins of his Winesaps, salvaged any young trees that looked like they might live, and replanted his orchard. Six years later, when the trees matured, one of them began producing sweet, mild, golden apples months earlier than the red Winesaps. Harvey realized he had a volunteer, and a volunteer with vast potential. The new apple had a lovely crispness and ripened crazy early, long before any other crisp apples. He named the apple for his wife (though I think he should have gone with the hurricane), promoted it enthusiastically, and saw his baby become a success. Ginger Gold has staked out a strong position as the national go-to apple for summer. It's a bit freakish that any apple could ripen so early and still be so crisp, and apple breeders would do well to delve into Ginger Gold's genes for the secret.

It's amazing how often stores and farmstands mention the (nonexistent) ginger note in this apple, not knowing it was named for a redhead. The apple has no notable flavor of ginger or anything else interesting, though its reliably sweet, crisp, and mildly tart profile makes it very useful, especially in salads, where its nonbrowning nature is a big plus. (Apples contain enzymes that react with oxygen, when the flesh is exposed to air, to produce harmless brown compounds, but certain apples lack these enzymes.)

Gravenstein

Alias Grav **Origin** Denmark (or possibly Italy), 1600s. **Appearance** Looks like a child's drawing of a fire, with jags of coppery flame licking up the sides against a yellow-orange glow—an effect enhanced by the waxy skin, which after just a couple of weeks in storage can feel as if it's been dipped in bacon fat. This is a large, gorgeous apple, with a hint of ribbing, particularly at the flower end. **Flavor** Famed for its high, fine, berrylike fragrances and sweet-tart taste, Gravenstein is juicier and sweeter than its late-summer cohorts. California Gravs don't have that much tartness, and northern Gravs do only if you pick them early. **Texture** As the English horticulturalist Edward Bunyard put it, "The flesh contrives to be crisp without hardness." An admirable feat, but it doesn't last; the apple bruises if you even think hard thoughts about it. Mush is not far behind, with the agreeably thin skin getting downright greasy. **Season** The first great apple of the year, the Gravenstein is an event in Sonoma County, California, every August, filling local markets and powering its own fair. In the Northeast, it ripens in late August to early September. **Use** A sublime treat straight off the tree in August. People rave about Gravensteins in pies, but man, you'd better make those pies the day you pick 'em. Great for sauce and first-of-the-season fresh cider. Superior for drying. **Region** Sonoma County is Gravenstein Central, but it is also popular in the Pacific Northwest, New England, Scandinavia, and Nova Scotia.

Legend has it that the first Gravenstein sprouted in the town of Gråsten (or Gravenstein in German), Denmark, near the German border, around 1669—though others claim that the 1669 tree was brought from Italy as a gift to the Duke of Gravenstein. It became a hugely popular apple in Germany, Sweden, and Denmark (which in 2005 claimed it as its national apple). It was first imported to North America in the 1790s, where it was soon considered the choicest of the summer apples, but it never caught on big in the East. In California, however, it arrived in 1812 with Russian immigrants from the Black Sea city of Sevastopol (who named their new city after their old one). For more than a century, Gravenstein orchards sprawled elegantly over much of Sonoma County, until they were ripped out to make room for vineyards and trophy houses in the 1980s and 1990s. In 2002, Sonoma County Slow Foodies began a movement to save the Gravenstein, reviving the old Gravenstein Apple Fair in Sebastopol. In 2012, Slow Food USA boarded the Gravenstein onto the Ark of Taste, its "living catalog of delicious and distinctive foods facing extinction." Still, Gravenstein acreage has shrunk to about 750 in Sonoma County, an all-time low.

The old pomology books rave about Gravenstein, with *The Fruits and Fruit Trees of America* (1845) calling it "one of the finest apples of the North of Europe" and *The Apples of New York* (1905) pronouncing it "perhaps unexcelled by any variety of its season." The esteemed twentieth-century California plant breeder Luther Burbank declared, "If the Gravenstein could be had through the year, no other apple would need be grown," while England's Edward Bunyard (1929), usually stingy with his praise, confessed, "Of Gravenstein it is hard to speak in mere prose, so distinct in flavor is it, Cox itself not standing more solitary, so full of juice and scented with the very attar of apple...bringing to mind the autumnal orchard in mellow sunlight."

Most summer apples are mealy, so part of Gravenstein's claim to fame is seasonal: crisp and cidery, it is the herald of the coming fall, a reminder that the crushing heat of summer will one day abate. Its distinctive flavor burns itself straight into the memory circuits and inspires legions of worshippers. One whiff of the apple can make German ex-pats weep with joy. Californians go all buttery for Gravensteins the way New Englanders do for Macs. I can relate. To me, the scent of Gravensteins summons up that childhood moment when your head is stuck in a pot of water with your face half-submerged, as you try to get your teeth into an apple and bring it to the surface before your older siblings drown you. Somewhere in that watery moment as you thrash around and try to trap an apple against the side of the pot, the better to sink your teeth into it, you inhale the essence of fall, and you never forget.

Red Astrachan

Alias Astrakan **Origin** Russia, pre-1816. **Appearance** Looks a lot like a Keepsake or a mini Gravenstein, gorgeous red mottling over yellow flame, with a dusty bloom when dead ripe. **Flavor** Tart and quincelike. Smells a bit like a Gravenstein. **Texture** Crisp enough to be pleasant when fresh off the tree, but softens with abandon once picked. **Season** June/July in warm regions, early August in the north. **Use** A summer thrill, good for sauce. **Region** A favorite of heirloom orchardists in Russia, northern Europe, and the United States.

Russia has developed thousands of apple varieties that thrive in brutal cold. In 1870, the United States Department of Agriculture decided it would be a good idea to bring some of those apples to the United States. Of the several hundred that were imported from the St. Petersburg Botanical Garden, Red Astrachan was one of the stars. (It had actually first landed on American shores in 1835, courtesy of the Massachusetts Horticultural Society, who got it from England, which had imported the tree from Sweden in 1816.) Not only did it prove it could grow in any climate, but in warm ones it bore crazy early—July in most of the United States, even June in the South. That helped it to become a hit throughout the United States, England, and northern Europe by 1900. With its bright red skin, it was a shockingly applelike presence in midsummer. Alas, its flavor would be considered applelike only when there are no true apples around for comparison. Red Astrachan is always tart and watery, with little sugar to balance the sour, and it rots exceptionally swiftly. It was always more for home use, though it once appeared throughout the South in every local market. There is about a nanosecond when it is still hanging on the tree and it gets its characteristic bloom, and in that nanosecond it can be a damn fine fruit. Before and after, make a lovely pink sauce out of it.

St. Lawrence

Origin St. Lawrence River Valley, early 1800s. **Appearance** Light-red ink blots dripping down over a yellow-orange canvas. Circus stripes, like a Barnum & Bailey's tent. **Flavor** Mild, sweet, low acid. **Texture** Always seems a little soft and disappointing. **Season** Early September, even late August if you want some tartness. **Use** A decent fresh-eating late-summer apple, a poor man's Gravenstein. **Region** Both sides of the St. Lawrence Seaway, from the Great Lakes to the Gaspé Peninsula.

Looks are the best thing about this apple, which has been well known in the regions bordering the 45th parallel for nearly two centuries, despite the fact that no one loves it. A Vermont horticulturist writing in 1901 said that in the Lake Champlain region "it is rather common but not highly prized," and that pretty much sums it up. If you lived in the far north and needed an apple that would come ripe before Snow and Pomme Grise, St. Lawrence served your purposes. Like many a Vermonter, I have an old St. Lawrence tree in my backyard. Early every September, I admire its sporty globes of color; then I blitz all its apples into sauce.

Summer Rambo

Aliases Rambour Franc, Rambour d'Ete **Origin** Rambures, Picardy, France, early 1500s. **Appearance** A huge, flattened apple with the pockmarked red skin of a mango. Its lenticels break the skin's surface, like Braille. The eye is wide enough to stick a pen inside. **Flavor** Honeyed and aromatic, with a mysterious artichoke finish. **Texture** Like a crisp, crunchy root vegetable in July. Quite juicy. Tender and grainy in September. **Season** Pick it green in July for pies, or red in August, and eat it fast. **Use** Superb pie apple when green; good dessert apple when red. **Region** Rambo has conquered the world; you'll find it in all temperate zones around the globe.

When I was a kid, there were a few years in the 1980s when "Summer Rambo" was a regular event at the multiplex, and in my mind I can't shake this ripped apple's unfortunate association with Sylvester Stallone. The overbearing presence bordering on self-parody, the cheesy skin, the prominent ribs, the beady eye. But the Summer Rambo predates John Rambo by a good 350 years, having first been recorded in 1535 in the town of Rambures, near the Somme River in northern France, and it has been a favorite on the Continent all that time. It's a favorite of many American apple collectors, too (like Tom Burford, author of *Apples of North America*, who ranks it among his Top 20 Dessert Apples). Yet it never appealed to commercial growers, and in 1907, the farming magazine *The American Thresherman* reminisced about the already disappearing fruit: "A boy would go farther to swipe Rambo apples, and subject his pantaloons to greater exposure from ugly dogs, than he would for any other kind, and boys know on which tree the best apples grow. A drink of cider without any fixin', made of Rambo apples, will go farther down and awake the molecules of mankind in a greater degree than any other kind of cider. The world is growing wiser, but not in raising Rambo apples."

Why isn't Summer Rambo a blockbuster? Many people seem to have a dim opinion of this apple because they don't pay attention to it until September, when it is already past its prime. Picked in July, while still fully green and tart, it makes a celebratory first apple pie of the year. As soon as it turns red, it's ready for eating fresh.

One person who is a fan is the author David Morrell, who in 1968 was struggling to find a name for the main character of *First Blood*, his novel-in-progress about a Vietnam vet. As he cogitated, his wife came home with apples from a local farmstand. He bit into one, loved it, and asked what it was called. "Rambo," she said. Morrell rushed to his typewriter, and an action icon was born.

Yellow Transparent

Aliases Early Transparent, June Transparent, White Transparent **Origin** Russia, 1800s. **Appearance** A large, round, bright yellow beauty with porcelain skin and bubbly green dots. **Flavor** Light and lemony. Quite refreshing. Never overly sweet. **Texture** Tender, tender, tender. **Season** June to late July, depending on region. **Use** Eat fresh before it knows it's been picked. Superb in sauce. **Region** Very popular in Europe, the rural South, and in nursery catalogs across the United States.

It's shocking to bite into a ripe Yellow Transparent and watch it turn brown before your eyes. This thing was not built to last. If you pick it a little green, it will keep in the fridge for a week or so, but then it will indeed be tart. Its sugar comes only right at the end, at which point you need to eat it right away, because it will turn dry and mealy fast.

None of this sounds very appealing, but Yellow Transparent is one of my favorite apples. It can be so rewarding. I have a small tree in my back field, and its apples look pretty much like all the other apples until some point in July, when all of a sudden they take off like they've been injected with growth hormones. In just a couple of weeks, they balloon into big, gorgeous, delicate yellow apples, and apples have been so far off my radar here in berry season that it takes me a day or two to realize that, by golly, those things are ripe. And I know they don't last, so there is always a little orgy of Yellow Transparent picking and eating, and when I get them just right, they have a Jolly Rancher scent and just enough sweetness and *snik* to be truly delightful. If you don't get them just right, do not despair; they melt down into some of the smoothest sauce you'll ever see. Transparents bear early and often, which adds to their generous nature.

Yellow Transparent was the most successful of all the Russian varieties brought over by the USDA from St. Petersburg in 1870. It was soon offered coast to coast. Many a southerner grew up sneaking Transparents from the tree in the front yard, and many a tree stands there still.

DESSERT APPLES

A dessert apple is any apple eaten fresh, on its own, out of hand or sliced, a good eating apple. It need not be eaten for dessert; the term comes from the old English idea of serving fresh fruit at the end of a meal. For example, Edward Bunyard's 1929 classic *The Anatomy of Dessert* is entirely devoted to fruit (plus a few words on wine).

Dessert apples perform solo and unplugged, with no backing from spices, ovens, or presses. Any apple that can hold your attention during the time it takes to devour it to the core is a great apple indeed, but historically there has been no consensus about what makes a great dessert apple. Some liked their fruit marrow-tender, while others wanted as much crunch as possible.

Thoreau preferred apples with "a kind of bow-arrow tang," while Asia, far and away the largest apple market in the world, likes its apples sweet and tropical. One thing all agree on is that the skin must be thin. A covering that adds a little crack and snap to the undertaking is a pleasure, while a thick, chewy hide ruins the whole experience. The nubby peel of a russet apple offers a whole different take on the pleasures of an apple.

Unfortunately, we are creeping disturbingly close to consensus on what an apple should be. Since Honeycrisp burst onto the scene a decade ago, most of the apples selected for commercial cultivation are cast in its image. All these apples are unquestionably great, a big improvement on many heirlooms (some of which survived due only to lack of competition in their little corner of the world), yet many are also indistinguishable from one another. A bit of a Hollywood blockbuster phenomenon is at work. The only apples making it out of the production studios into worldwide release are the ones that fit the formula: lots of crisp explosions, with a sweet ending. As consumers learn to expect this from every apple, they may forget how to savor the quirky or nuanced ones. The following apples run the gamut, from explosively crisp to spicy-tender. Together, they offer you a garden full of experiences unmediated by any focus group.

Ambrosia

Origin Similkameen Valley, British Columbia, 1990s. **Appearance** Beautiful curtains of carmine drape over a creamy yellow background. Ambrosia's five ribs are prominent, and it sits on five little knobs. The flesh is dreamy. **Flavor** You can already smell the banana as you raise the uncut apple to your mouth. Once you break into the flesh, banana and pear esters surround you. The apple is mildly sweet, passingly tart, and tropical, like one of those modern fusion juices. Very refreshing, not overwhelming, but subtly sublime. **Texture** The epitome of a modern apple, Ambrosia's crisp and breaking flesh sprays juice in all directions. It is a bit more tender than the explosively crisp apples. **Season** September. **Use** An ideal in-hand snack. Also excellent for salads because of its nonbrowning qualities. **Region** Pacific Northwest, but coming on strong elsewhere.

Ambrosia fits seamlessly into the line of what I think of as Pacific Rim apples: tropical and sweet profiles, with little or no acidity, that appeal to the Asian palate that dominates twenty-first-century apple consumption, with Fuji and Gala being the standard-bearers. It also pulls off what is called the bicolored look, which is in vogue in these post–Red Delicious times. Combined with its recent birth, this makes Ambrosia the model of a modern major apple. Yet Ambrosia also gives us hope that the glorious world of spontaneous apple evolution is still alive. Over the past half century, only a handful of new apples have not been the product of a university breeding program, yet Ambrosia is a throwback, a chance seedling that snuck into a new planting of Jonagolds on the British Columbia orchard of Wilfred and Sally Mennell. Really, it was first spotted by the pickers, who would pass up the Jonagolds and strip clean the mystery tree for their own use. The Mennells trademarked the apple and began selling it into the organic market, where it was an immediate hit. Fun and uncomplicated, and so darned pretty, Ambrosia is the perfect summer fling apple. You might not remember it years from now, but you won't get hurt, either.

Autumn Crisp

Alias NY674 **Origin** Geneva, New York, 2009. (Cross of Golden Delicious and Monroe.) **Appearance** Medium size, classic apple profile, a brilliant pink-tinted red over yellow background. As waxy as a Corvette hood. **Flavor** Pleasingly sweet-tart, but somewhat watery. **Texture** The same unlikely explosiveness you find in the Honeycrisp. Snaps like a chip. Melt-in-your-mouth flesh. Dissolves into juice at the slightest provocation. **Season** Early fall. **Use** Eat out of hand. Perfect salad apple: light, crisp, and nonbrowning. **Region** New York State; uncommon.

A new apple from the storied breeding program at Cornell, which brought you the Cortland and the Macoun. This is the old guard of the apple world, which in recent times has tended to give its apples bombastic American names like Liberty, Freedom, and Empire. Meanwhile, the upstarts at the University of Minnesota apple breeding program seem to name theirs after strippers: Sweet Sixteen, Frostbite, Honeycrisp, SweeTango. Guess whose apples get all the press? Well, in 2009 the Cornell team finally got on the bandwagon with Autumn Crisp (and will pile on in 2015 with SnapDragon and RubyFrost). The apple is a dead ringer for Honeycrisp: its large cells shatter when you bite into it, spraying juice everywhere and leaving you with a mouthful of apple water. It is wicked fun to bite into, and utterly forgotten thirty seconds later, which seems to be the unchallenging direction modern apples are headed. There are a lot of points in this apple's favor: it's beautiful, kids love the crispness and gentle flavor, and it takes forever to brown after cutting. If only it didn't try so hard to please.

Baldwin

Aliases Woodpecker, Pecker, Butters **Origin** Wilmington, Massachusetts, 1740. **Appearance** Emphatically red on the sunny side, light red elsewhere, with big, indented white dots. You can often distinguish Baldwin from other red apples by the brick color it displays at the edges of the red, especially on the shoulders, a strange hue the color of a green cherry tomato that is beginning to turn orange. Ours pictured here is a bit on the young side, its red just beginning to develop. The apple can be large and heavy. Its white flesh is sometimes marred by brown flecks known as Baldwin Spot. **Flavor** In fall it is plenty tart and, in my experience, a little weak in flavor, with a sort of savory olive note to the skin, but in November it mellows and becomes, to many people, the quintessential sweet-tart apple. **Texture** Very hard and substantial. The supremely thick skin made it popular for storage and shipping. **Season** Pick in October. **Use** People have raved about Baldwin's fresh eating qualities for centuries, but to me it's even more successful in pies (where its hardness holds), when dried (a mysterious chanterelle flavor emerges), and in hard cider (West County Cider in Massachusetts makes one using apples from century-old trees that is light and crisp with a good green strawberry finish). **Region** Still common in New England, if not abundant. Fails miserably in warmer climates.

The nineteenth century was the century of the apple in America, and Baldwin was the apple of the nineteenth century, at least in the northern states. (It thrives as far south as Missouri and the Midatlantic, but produces its tastiest fruit in climates like that of its native Massachusetts.) Baldwin was what baseball analysts would call a five-tool player: The tree cranked out huge quantities of apples, was impressively disease-resistant, and the handsome, red apple was delicious, long-lasting, and made fantastic hard cider. If you were setting up your self-sufficient homestead, Baldwin was the first apple you'd plant.

And plant it people did. The original tree was spotted by William Butters on land owned by his kinsman James Butters (and which still abuts Butters Row, where a granite monument now commemorates the spot of the original tree) in Wilmington, Massachusetts, in the 1740s. Butters transplanted the tree to his farm, and it became briefly known as the Butters apple, but around 1780, a surveyor from neighboring Woburn named Samuel Thompson spotted the gorgeous red fruit while working in the area. According to the account of Thompson's grandson, "The tree was hollow with decay, and a woodpecker had found a place for her nest therein. He said he carried home some of the fruit and gave his brother Abijah some of it, and they were so highly pleased with it that they procured a lot of scions from the tree and set them in the trees around their homes, and they soon began to bear fruit." Thompson named the apple the Woodpecker, "after the old bird," which was soon shortened to Pecker, the name under which the apple became much celebrated in Woburn until the Thompsons' neighbor, Colonel Loammi Baldwin, became greatly enamored of the apple, "and he valued them so highly he went into them deeply and spread them around among his friends broadcast, and they had no name for them, and of course they gave it his name."

By the early 1800s, Baldwin had become Bostonians' favorite apple, and by 1850 it was the dominant apple in New England and much of New York. In 1905, Spencer Ambrose Beach wrote in *The Apples of New York*, "Probably more Baldwin apples are put upon the market than all other kinds in the state put together." Further, it was "preeminently the leading variety in the commercial orchards in New York, New England, certain regions in Southern Canada, in the southern peninsula of Michigan and on the clay soils of Northern Ohio."

Yet for all its talents, Baldwin had two flaws that would prove its undoing. The first was a strong biennial streak. Apple trees bear fruit on second-year growth: a spot on the tree that fruits one year will send out new spurs the following year, which will fruit the year after that. For this reason, many apple trees have a tendency toward biennial bearing—after a heavy fruit year, they will bear little or no fruit while they focus on new growth. Baldwins are one of the worst offenders. This is a minor inconvenience in a home orchard, but it can ruin a commercial enterprise, where growers are forced to deal with a glut of fruit one year and almost none the next. So even though Baldwins were in many ways the first perfect commercial apple—remarkably uniform in size and shape, bright red (which was prized even in the 1800s), and excellent shippers—horticulturalists were already looking for a new and improved apple when Baldwin's second flaw expedited things. Baldwin had always been susceptible to "winter kill," and then came the infamous cold snap of February 1934. A winter thaw had triggered some sap flow in the trees, then temperatures plunged to 40 below zero, and the Baldwins literally exploded as the sap inside them froze and expanded. Millions of trees died and numerous orchards were ruined. For Baldwin's successor, agricultural extension agents looked north, where an extremely cold-hardy variety from Canada was already winning converts. Not only could the McIntosh survive an ice age, it also tended to produce a perfect set of fruit every year. Few Baldwins have been planted since.

Blenheim Orange

Aliases Blenheim, Blenheim Pippin, Kempster's Pippin, Goldrenette von Blenheim **Origin** Woodstock, Oxfordshire, England, 1740. **Appearance** Very large and obviously oblate, with smooth green skin, turning yellow when ripe, covered by a charming orange-red blush on the exposed side. **Flavor** Sweet and zingy when young, with the tartness fading by November to reveal an intense nuttiness. **Texture** Coarse grained, not terribly juicy, Blenheim Orange eats like a russet. **Season** Ready for pies in September and fresh eating from October through December. **Use** Great in pies, sauce, cider, and out of hand. Keeps well, too. **Region** England, where it is well known; grown by collectors in the United States.

A famed English apple for both the kitchen and out-of-hand eating, the first Blenheim Orange was planted by a tailor named George Kempster near Blenheim Palace, the monumental and architecturally dubious "country house" built near Oxford as a gift to the Duke of Marlborough, who had just defeated the French and Germans at the Battle of Blenheim in Bavaria in 1704. (You've seen Blenheim Palace in a zillion films, from *Harry Potter and the Order of the Phoenix* to *Hamlet*.) The apple was known locally as Kempster's Pippin (a pip being a seed, and a pippin a seedling), but once it achieved fame and fortune in London around 1820 it received a more regal rechristening. It should be pronounced "Blennem," spoken as if you were Winston Churchill, whose family has resided in Blenheim Palace for three hundred years.

Of the huge apples that lend themselves so well to pie-making, few have Blenheim's rich nuttiness. Your pie will have an extra dimension if you include a Blenheim, and, because Blenheims can be quite tart in their youth, a straight-Blenheim pie is a fine thing, too—though Edward Bunyard, the 1920s British gourmet, would be horrified if you did anything with a Blenheim but eat it fresh. To him, it held "a nutty, warm aroma which is to my taste the real apple gust." If you don't cotton to nutty apples, well, Bunyard probably wouldn't cotton to you. "The man who cannot appreciate a Blenheim has not come to years of gustatory discretion; he probably drinks sparkling muscatelle."

Blue Pearmain

Origin New England, probably Boston region, late 1700s or early 1800s. **Appearance** Stunning. In early fall, Blue Pearmain's matte finish is mottled with pink, green, yellow, orange, taupe, auburn, carmine, and black, all airbrushed with the loveliest powdery blue bloom. Often there is a burst of russet out of the stem end. By October a cold-initiated wave of reddish purple overtakes this large apple and masks the other colors. There will be stripes of near black on the sun-exposed side and blurry white lenticels. Around the flower end, the purplish color disappears and it looks cracked and russeted, like the eye of some ancient sea beast. **Flavor** In September, a big burst of musky orange, flanked by a hard-charging kiwi brightness and a tropical melon finish. Later in the fall, the acid fades and the flavor is more mild and melonlike. **Texture** Quite firm, dense, a bit coarse, and not especially juicy. One Blue Pearmain makes a meal in itself. The skin is a thick canvas. **Season** Ripe in October (September in the South). Will keep through December, but the flavor gets less interesting the longer it is stored. **Use** Good out of hand, ideal for whole baked apples (see page 298), excellent for pies if used before the acidity fades. The thick skin breaks down surprisingly easily, making it a good candidate for a quick emergency applesauce. **Region** Heirloom orchards throughout the Northeast, and occasionally farther south.

Henry David Thoreau was not a fan of cultivated apples. "I have no faith in the selected lists of pomological gentlemen," he wrote in "Wild Apples," an essay he completed for the *Atlantic Monthly* as he lay dying of tuberculosis in 1862. "Their 'Favorites' and 'Non-suches' and 'Seek-no-farthers,' when I have fruited them, commonly turn out very tame and forgettable. They are eaten with comparatively little zest, and have no real tang nor smack to them." He preferred the "racy and wild American flavors" of seedling trees, with one exception: "I know a Blue-Pearmain tree, growing within the edge of a swamp, almost as good as wild." He loved to gather apples from it, and though he doesn't specify why, it may have had something to do with his admiration for the way that apples, after a hard frost, turn color "as if accidentally sprinkled from the brush of him who paints the autumn leaves." Blue Pearmain feels layered, as if its skin were a canvas worked over by a patient and exacting Impressionist. You can lose an hour gazing into its textured depths. It tends to feature raised russeting like old Celtic knot patterns, which gives Blue Pearmain the feel of some ancient talisman holding mysterious power. This is the apple Elrond would have tended in his backyard in Rivendell, and it would have been off-limits to any dwarf or hobbit.

Blue Pearmain loves the cold, which made it popular on hardscrabble New England farms throughout the 1800s. Although it's generally considered a northern apple, I've had Blue Pearmains straight off the tree in Virginia that were among my most sublime apple experiences. In fact, considering the disdain Beach shows for it ("Not a reliable cropper. It is apt to have a pretty high percentage of unmarketable fruit. The fruit is of mild flavor and does not rank high in quality. The skin is thick."), and that it is on the Virginia pomologist Tom Burford's list of his Top 20 Dessert Apples, I wonder whether this is one of those New England apples that might actually find its true self down south.

Braeburn

Origin Waiwhero, New Zealand, 1952. Appearance Streaky orange-red stripes, turning solid red with white dots, elongated and conical, acutely five-sided, this is a generically handsome apple. Flavor Concentrated, spicy cider sweetness balanced by zesty tartness. Texture Superb crispness, with skin so squeaky it gives me goose bumps. Its breaking, juicy flesh snaps off in eminently satisfying chunks. Season Early winter for U.S. fruit, midsummer for New Zealand and Chilean fruit. Use Best for fresh eating, decent as a cooking apple. Region Grown worldwide.

Think back to those dark days of the 1980s, when the supermarket was ruled by three apples: Red Delicious, Golden Delicious, Granny Smith. Monochromes of red, yellow, and green. Suddenly, an exotic stranger arrived in their midst, and it was bicolored. Those who dared taste the strange apple discovered that it had something missing from the Big Three: flavor. It was the Braeburn, and it blazed the trail for acceptance of other bicolors like Fuji and Gala.

A Braeburn is a reliably satisfying eating experience and a supermarket manager's dream. Unlike some of the other modern apples, which seem to fancy themselves crunchy mangos, Braeburns are emphatically appley. Odd that such fine apples, whose U.S. production has been flat for a decade, seem to be losing traction to some of the newer kids on the block.

Burgundy

Origin Geneva, New York, 1953. (Cross of Macoun and Antonovka.) **Appearance** Like a giant black cherry. **Flavor** Surprisingly light and refreshing, like tart, watered-down apple juice. **Texture** Snappy and juicy in early September, it goes downhill fast and becomes quite soft. **Season** Early September. **Use** Eat as fresh as possible. **Region** Northern farmstands and pick-your-own orchards.

The name Burgundy doesn't do this apple justice. Cabernet might have been better. It can give Red Delicious and Arkansas Black a run for their money in the midnight-red department. In many ways, it is the platonic ideal of appleness, the apple that pops into your head when somebody utters the word: round, slightly flattened, and a glossy, unforgettable red. When Cornell's New York State Agricultural Experiment Station released this apple to the industry in 1974, it included a funny comment on the color in its notes: "Its very dark color may be too dark in the opinion of some people. However, limited studies indicate no consumer rejection due to its dark color." Far from it. A bowl of Burgundies is a beautiful thing to behold. The flavor is middle-of-the-road apple—Jonathan and Wealthy come to mind—and the apple doesn't keep at all. The NYSAES intended it for early September farmstand sales, since it comes ripe two or three weeks before McIntosh and the other big varieties, and that is still where Burgundy thrives best. By the time the good stuff arrives in late September, Burgundy is long forgotten.

Chestnut Crab

Alias Chestnut **Origin** St. Paul, Minnesota, 1946. (Cross of Malinda and a crab apple.) **Appearance** Like one of those little doughnut peaches, a flattened swirl of orange, red, and yellow, covered with a fine, papery russeting. **Flavor** Like a peach pie on a graham cracker crust. **Texture** Uniquely fine grained and snappy without being hard. Perfect, bite-size chunks snap off in your mouth and disintegrate as you chew. **Season** Ripens in September, doesn't last past October. **Use** The perfect fall dessert. Cooking it would be a shame. **Region** Nationwide; rare.

One of my favorite apples. The Chestnut Crab looks more like a little peach than an apple and tastes somewhere in between, with a baked cookie quality to boot. (This turns to a chestnut flavor once the apple is past its prime.) It is very, very sweet, but with lots of aromatics (nutmeg! Coca-Cola!) and enough acids to keep things from getting cloying. Ah, the nutty things the apple genome is capable of. I love Chestnut Crab's tawny skin, with that autumn-leaf palette and texture, love to rub it against my lips and feel that dry, papery skin. There is a Bronze Age ancientness to it, a cradle-of-civilization, ribbed-fruit feel to the thing, that makes me think of fine parchment and old maps. Bite into one and you are Captain Sir Richard Francis Burton, venturing in disguise to exotic lands. The natives have just handed you a mysterious fruit, and you snack on it and marvel at its strangeness and think, "The world is full of surprises."

And it's all a fantasy, of course. The Chestnut Crab hails from the exotic locale of St. Paul, Minnesota, where the University of Minnesota crossed Malinda, the foster mother of the Honeycrisp, with a crab apple back in 1946 to create this little wonder. Nothing they have done in the ensuing sixty-eight years has produced anything half as fine, which makes it perplexing that the Chestnut Crab remains virtually unknown. That may soon change. Apple geeks tend to obsess over the little Chestnut, planning their Septembers around its arrival, and apple geeks are on the rise.

Claygate Pearmain

Origin Claygate, Surrey, England, 1822. **Appearance** This little chameleon can be gold, green, or red, smooth skinned or russeted, depending on where it is grown and when in the season you find it. Typically, it will be a matte gold, like our model here, with some red streaking on the sunny side. **Flavor** Deliciously spicy, the Claygate opens with fruity pineapple sweetness, then blossoms with watermelon rind or green tomato aromas as you chew (and chew you must!). The finish echoes with notes of *guarapo*, pear, and nutmeg. **Texture** Remarkably dense, a Claygate Pearmain is a workout for your jaw. You sense the density as soon as you pick it up. It feels like a squash ball in your palm. When fully ripe, that crunchiness mellows into delightfully tender, juicy quanta of happiness. **Season** Ripe in October, peaks in flavor in November and December. **Use** Best eaten fresh, but a decent cooker, too. **Region** Revered in the United Kingdom, but still rare in America.

Claygate is one of those astonishingly aromatic British varieties of apple that haven't made nearly the inroads in America that they should have. The whole, irreducible essence of the thing reminds me of a Painkiller, the classic Caribbean rum drink, or even (don't call me crazy until you've tried it) a traditional Japanese kaiseki breakfast, sprinkled with toasted nori flakes. Complex, surprising, refined, and balanced. *The Book of Apples* describes it as having "a definite taste of walnuts." The original tree was spotted in a hedge in the village of Claygate, and went on to a modicum of fame in Victorian England. In 1929, Edward Bunyard described it as "a comfortable-looking fruit in sober russet with a tender granular flesh which has much of the Ribston richness, and is fully deserving of a place in the best dozen dessert apples." I concur.

Cox's Orange Pippin

Origin Colnbrook, England, 1825. **Appearance** A yellow background spackled with a little red and a lot of orange in a rustic-looking patina that looks like it came straight off the Sistine Chapel (*before* the cleaning). The flesh has the cream-yellow tinge of most rich-flavored apples. **Flavor** Think ambrosia salad—pineapple, oranges, marshmallow, and coconut. With lime squeezed over the top. And a sprinkle of flower petals. Cox rarely loses a taste contest. **Texture** The breaking flesh sprays juice upon every bite. With a perfect mix of crisp and tender, and nice thin skin, Cox goes down easy. **Season** Fresh picked in September, it is delicious but tart; by late October, it has mellowed into its full tropical apogee. After that, it goes downhill fast. **Use** Cox would make a fine pie or sauce, but this would be like using the *Mona Lisa* as kindling. **Region** A staple of UK supermarkets, though often in subpar condition there. Look for it at roadside stands in England, the Canadian Maritimes, New England, the Finger Lakes, the Great Lakes, and the Pacific Northwest. Many feel that Cox only sings in a maritime clime.

I ate my first Cox while driving through the Finger Lakes region of New York. Sitting beside me on the passenger seat was a bag of Cox's Orange Pippins I had just picked up at a farmstand. I grabbed an apple, bit into it, and my world changed. A kaleidoscope of fruity esters burst across my sinuses. Visions of colorful fruits spun in my eyes like in a slot machine. I had never tasted anything like it. There was lots of mandarin in there. It was sprightly, exotic, spicy... my brain strained for adjectives. It was like cherry-vanilla ice cream with lemon zest and rose petals. It was heavenly. What it was not was *appley*. Those flavors that I'd always thought defined an apple—that generic fruit-juice flavor of Golden Delicious and McDonald's apple pie—played little if any role in Cox. I wasn't sure I understood what "apple" meant anymore. Cox's complex, nuanced, poignant character can hold a room spellbound, yet it has also delivered too many lackluster performances. This finicky apple's greatness must be coddled.

It is happiest in its native England. Ask someone from Britain to name their favorite apple, and they're likely to pipe "Cox's Orange Pippin!"—a tree originally planted from a Ribston Pippin seed in the village of Colnbrook (near the current location of Heathrow Airport) by retired brewer Richard Cox. Victorian England went gaga for the apple's exotic, spicy-citrus aromatics, but the darn tree never met a disease it didn't like contracting, and commercial growers had widely given up on it until lime sulfur sprays came to the rescue in the 1920s, allowing Edward Bunyard to speak casually of the "Coxomaniacs" who planned their year around the arrival of this apple. It is still the UK's number-one apple, and still a devil to grow organically.

A not insignificant portion of the apple-breeding world has devoted itself to transferring those Cox flavors into a less finicky package, with mixed results. Like a prize stud bull, Cox has been used by breeders, both amateur and professional, to sire at least eighty successful children, including Ellison's Orange, Elstar, Fiesta, Freyberg, Golden Nugget, Holstein, James Grieve, Karmijn de Sonneville, Kidd's Orange Red, Rosy Blenheim, SunCrisp, and Tydeman's Late Orange. Its grandchildren include Gala, which is to Cox as Drew Barrymore is to John Barrymore.

Empire

Origin Geneva, New York, 1966. (Cross of McIntosh and Red Delicious.) **Appearance** A medium apple with deep, shiny red skin and a purple bloom, this Macoun lookalike is a greengrocer's dream. **Flavor** Sweet and softly twangy. In *The Apple Lover's Cookbook*, Amy Traverso describes Empire as tasting like "rosehip tea with honey," which is so spot-on that there's no need for anyone to ever attempt to describe this variety again. **Texture** Straight off the tree, it has some snap to it, but it doesn't last long. **Season** Ripe in September. **Use** Eat fresh in September, when it is still firm and a little tart. Softens up if not in controlled storage. **Region** Extremely popular in New York State.

Empire combines the fire-truck sheen of a Red Delicious with a zippier flavor. It's still a middle-of-the-road sweetness, but the overall package has been enough to vault Empire into the top ten sellers in the United States in recent years. A good multiuse apple that excels at nothing, except looking good and cranking out heavy crops every year—which, any apple grower will tell you, is certainly not nothing.

Esopus Spitzenberg

Aliases Spitz, Spitzenberg, Spitzenburgh **Origin** Esopus, New York (Ulster County), 1700s. **Appearance** The red coat, spotted with fat yellow lenticels like bubbles in hot oil, makes me think of the hide of some reptilian beast. The greenish yellow background often shows through. Modestly sized and modestly ribbed. When it ripens fully, the flesh turns yolk-yellow.
Flavor Extraordinary brandied, burnt-sugar notes, as if it were already halfway to being a tarte Tatin. One thinks of burnt orange, or Cointreau, and (after some mellowing) lychee and roses. **Texture** Excellent breaking crispness. Quite juicy. The skin is on the chewy side. **Season** Pick in October. Give it a month to mellow unless you are a sour fanatic. The flavor takes off around the winter solstice, and will hold in cold storage through spring. **Use** This apple can do anything. Eat it fresh, make it into a pie, ferment it into an unforgettable cider. **Region** The new darling of home orchardists and cider makers throughout the Northeast and upper Midwest.

"Spitz" is the one American apple that can go up against the Cox's Orange Pippins and Orleans Reinettes of the world. (And, like Cox, it shows a disheartening tendency to succumb to every disease that wafts through the orchard, and even when healthy bears only lightly.) Once Thomas Jefferson tasted it, he determined to make it a big part of his orchard at Monticello. (The Virginia climate had other ideas.) Word on the street is that it's President Obama's fave, too. This apple has the haze of greatness hanging around it.

And yet, if you taste it too early, just off the tree, you wouldn't know it. Spitz pushes both the sour and sweet about as far as you can push it. It's aggressive. So acid, in fact, that when Steve Wood of Farnum Hill Ciders began working with it, he thought his hygrometer was broken. It told him that the Spitz juice was sky-high in sugar, but he couldn't taste it through the acidity. Spitz is also full of wheatgrass notes when still green. Once it mellows, however, after a month or two, the acidity settles down and the flavor blossoms. Upon biting into his first Spitz, my thirteen-year-old son said, "Wow! Lychee soda!"

Spitz can shine in a single-variety hard cider—wildly tart, with aromas of Juicy Fruit gum, anise, and roses—and is even better when teamed with less acid apples.

Yet for all its flashes of greatness, Spitz has never quite put together the dominant career it should have. It was already heralded in 1817, when Coxe praised its "exquisite flavour." In his famous 1850s sermon on apple pie, Henry Ward Beecher asked, "Who would put into a pie any apple but *Spitzenberg*, that had *that*?" In 1905, Beach didn't beat around the bush, calling it "the standard of excellence for apples of the Baldwin class... unexcelled in flavor and quality." Yet the light crops it bore kept Spitz out of the commercial orchards, and it took a beating during the twentieth century. Fortunately, that dark period is over, and Spitz is once again the phenom that can't miss.

Fuji

Origin Fujisaki, Japan, 1939. (Cross of Ralls Janet and Red Delicious.) **Appearance** Big and barrel-shaped, with pink stripes over a yellow background, turning brick red on the sunny side, but always striped. **Flavor** Unique. A colossal opening salvo of lychee gives way to tantalizing touches of mango, cantaloupe, and gourmet jelly bean (choose any flavor you want; they all taste the same). **Texture** A crisp fountain of juice. Skin can be chewy. **Season** October, but it stores quite well and is grown everywhere, so a quality Fuji is never out of arm's reach. **Use** Eat fresh. Also makes an amazing dried apple; the gourmet jelly bean really shows through. **Region** World domination has been achieved. Fujis thrive in warm climates, but those from cooler regions are more likely to have a refreshing shot of acidity.

After the Braeburn breached the walls of the U.S. apple market, opening grocers and consumers to the invasion of exotic, striped apples, Fuji came charging through the gap and sacked Red Delicious once and for all. Its off-the-charts sweetness, concentrated Asian tropical fruit flavors, and showers of juice were like nothing that had been seen before, and by the 1980s Washington State orchards were replacing their Red Delicious with Fuji as fast as they could. True, Fujis are harder to grow than the weedlike Red Delicious, but they brought two and three times the price. Fuji apples had already conquered the Japanese market in the 1960s (indeed, they were developed in Japan specifically for the domestic market), and today they have a virtual monopoly. One of the classic sights in Tokyo is of flawless, beautifully packaged Fuji apples selling for $10 or $20 apiece.

Apple people love to knock Fujis because their sweetness is not balanced by any real acidity, but in general the world does not have this issue. The world *loves* sweet apples. You, apple geek, may prize the tart ones, but you are an aberration. This is why Red Delicious and Golden Delicious ruled the roost in the United States until something even sweeter came along. This is why more than 70 percent of the apples grown in China are Fuji—and China produces about half the world's apples, eight times that of the second-place United States. There are five times more Fujis in China than there are apples in the United States. We tend to still put Red Delicious at the center of our pomological cosmology, but outside the United States, everyone knows that we live on Planet Fuji.

Gala

Origin New Zealand, 1920s. (Cross of Kidd's Orange Red and Golden Delicious.) **Appearance** The epitome of the bicolored, modern apple. Red on one side, yellow on the other, with a continuum of pinkish orange stripes in between. Waxy skin and golden flesh complete the coiffed ensemble. **Flavor** A good Gala is glorious, with a sort of strawberry-yogurt intensity. But be careful; the acidity disappears in storage, exposing a juice-box blandness. **Texture** Both crisp and tender, Gala at its best has the modern Honeycrisp snap. Too often, however, it can be a bit soft. It is one of the juice sprayers. **Season** Available year-round (from six continents; no Antarctic Galas yet). **Use** Excellent fresh, solid in pies and tarts. **Region** The world is its playground.

Gala is like a jet-setting movie star, comfortable (and stylish) in almost any continent or role. You'll see it in New Zealand, in Asia, in the UK (where it is squeezing out its grandparent, Cox's Orange Pippin), and more and more in the United States, where it is now the number-two apple, its production having doubled in the past decade. It continues to steal the limelight from Red Delicious, Golden Delicious, and even Fuji, a tasty but less versatile fruit. Gala has Fuji's florals and sweetness and shelf life, but better acidity, and it's an easier apple to grow.

Gala began in New Zealand in the 1920s when famed orchardist J. H. Kidd crossed his own celebrated if oxymoronic Kidd's Orange Red (a cross of Cox's Orange Pippin and Red Delicious) with Golden Delicious. That's a lot of famous forebears. Gala hit New Zealand consumers in the 1960s, but didn't begin its American blitzkrieg until the 1990s.

I used to sneer at Galas as sweet, bland kid's apples; then I plucked one off a tree in Washington State's Yakima Valley, bit into it, and thought it was one of the best apples I'd eaten in my life. Hit it right, and Gala will deliver an unforgettable performance. Other times, it seems to be mailing it in, providing some light, sweet entertainment with no edge. A tasty Gala should look like Cox's Orange Pippin, all light and orange-yellow and stripey; a mostly red Gala has been kicking around in storage for months and will have lost its zippy Cox notes, or, worse, is a new sport—a mutant selected for increased redness, at the expense of flavor. Beware of these Galas, which will bear cheesy names like Royal Gala or Regal Gala; they are heading down the Red Delicious road to ruin.

Golden Delicious

Alias Mullins Yellow Seedling **Origin** Clay County, West Virginia, 1890. **Appearance** Should be large and modestly conic, with round shoulders and a cheerful yellow skin dotted with brown freckles. Too often, it is more of a pistachio green, picked early to improve shelf life. Russeting is a good sign, as is a touch of pink on one cheek. **Flavor** Not exactly complex, but darn good in an appley sort of way. Mostly sweet, it is about as tart as apple juice, but there is a touch of intrigue there, a whisper of melony complexity that has been teased out of Golden Delicious's many illustrious offspring. **Texture** Fresh off the tree, a properly grown Golden Delicious has a lovely breaking quality. Each chunk eagerly separates from the mother ship and hurls itself into your mouth, where your teeth can have their way with it. **Season** September to October. If it ain't yellow, don't buy it. Will last well into spring. **Use** Fresh eating; holds up well in pies (though it needs lemon juice). **Region** Nationwide, though the best are grown in warmer areas. Ubiquitous in supermarkets in the United States and Europe.

Golden Delicious sets the standard for feminine appleness, the blond, curvaceous, soft-shouldered distaff to Red Delicious's angular fire-truck masculinity. If you think of it as a lame supermarket apple, you may have been hanging out in the wrong places. Left to its own devices, a proper Golden Delicious will form a fine network of russeting, as well as the palest blush on one sunny cheek (as on the sun-ripened Virginian in our photo). It will look downright old-timey. It will also taste rich, sweet, and cheerful, unlike the woody supermarket ones picked early to ensure that they'll be hard enough to ship. They are at their most cheerful in Mid-Atlantic zones not too far from their West Virginia cradle, where they can be left on the tree until fully ripe.

The saga of the Golden Delicious began in 1891 on the Clay County, West Virginia, farm of L. L. Mullins. That was when Mullins sent his fifteen-year-old son, J. M., to mow the fields. In 1962, J. M. Mullins, then eighty-seven, recounted to the *Charleston Daily Mail* what happened that day: "I was swingin' away with the scythe when I came across a little apple tree that had grown about twenty inches tall. It was just a new little apple tree that had volunteered there. There wasn't another apple tree right close by anywhere. I thought to myself, 'Now young feller, I'll just leave you there,' and that's what I did. I mowed around it and on other occasions I mowed around it again and again, and it grew into a nice lookin' little apple tree and eventually it was a big tree and bore apples."

J. M.'s uncle, Anderson Mullins, later came to possess the farm, and around 1905 began noticing the extraordinary tree. The only popular yellow apple in the South at the time was the Grimes Golden, of which Mullins had several growing nearby (one of which was probably the parent). But this was no Grimes. It was much larger, crisper, and spicier in flavor. The tree outproduced all the other trees on the farm, and the apples held up beautifully right into spring.

By 1913, Mullins had decided that he had something extraordinary on his hands, so he mailed some samples of Mullins Yellow Seedling, as he called it, to Stark Bro's, the Missouri mail-order nursery that dominated the apple market at the time (and continues to thrive today). Stark Bro's had scored a monster hit with their Delicious apple in 1895, and Mullins thought maybe they could do something with his apple. In April 1914, he sent them three more apples—a cagey move on his part, because in that season the laudable keeping powers of the fruit were obvious.

Brothers Paul and Lloyd Stark were more interested in red apples, which had more market appeal, but when they tasted Mullins's apple they had an epiphany. "We had never experienced such a spicy flavor before,

especially from a yellow apple," Paul Stark later wrote. The main yellow apple of the time was Grimes Golden, but that apple's small size had always limited its popularity. Stark decided that a big, crisp yellow apple to complement their Red Delicious would be a fine idea, so he traveled a thousand miles by rail, and the final twenty-five by horseback, to reach Mullins's farm. No one was home, but Stark could see an orchard on the slope behind the house, and he began to poke around. Most of the trees he saw were in poor condition, and he must have begun to doubt that he could possibly have the right place. And then, as he later recollected, something caught his eye. "There, looming forth in the midst of small leafless barren trees, was one tree with rich green foliage as if it had been transplanted from the Garden of Eden. That tree's boughs were bending to the ground beneath a tremendous crop of great, glorious, glowing golden apples. I started for it on the run, a fear bothered me. Suppose it's just a Grimes Golden apple after all? I came closer and saw the apples were 50 percent larger than Grimes Golden. I picked one and bit into its crisp, tender, juice-laden flesh. Eureka! I had found it!"

Starks paid Mullins $5,000 for the propagation rights to the tree and for the nine hundred square feet of ground around it. He built a wood-and-wire cage around the tree, to discourage midnight grafters, complete with an electric alarm. In 1916, he introduced the fruit to the world as Golden Delicious, and it went on to fame and fortune, as well as a career as the Secretariat of the apple world, siring Jonagold, Ozark Gold, GoldRush, Mutsu, Arlet, Elstar, Pinova, Gala, Pink Lady, and many others. In fact, the prevalence of its genes in the world apple supply helped get it chosen as the apple to decode for the Apple Genome Project, which in 2010 published the complete sequence of Golden Delicious's genome.

For decades, Golden Delicious placed second in U.S. apple production, several lengths behind Red Delicious. Red Delicious held half the market, while Golden Delicious maintained a respectable 15 to 20 percent. But in 2006 Gala, its own child, bumped it down to third, and today Golden Delicious holds on to about 10 percent of the market. (In Europe, however, Golden Delicious has been the top apple since 1945, when it arrived as part of the Marshall Plan to reboot French agriculture, and it continues to hold about 25 percent of the market, more than twice that of any other variety.) What accounts for such popularity? No one, other than Paul Stark in full P. T. Barnum mode, has ever claimed it was a gustatory knockout. Rather, Golden Delicious is the apple that does everything well enough, while being a grower's dream. It makes a nice, big, pleasant, fairly crisp and very sweet apple with a middle-of-the-road flavor profile and good crispness, it holds up in pies, and it lasts for a long time in storage. For growers, it produced bumper crops each year with little drama. It was not a complicated apple.

That was an excellent formula for success in the 1950s. Even in the 1990s, the typical American supermarket would have three apples: green Granny Smith, yellow Golden Delicious, and red Red Delicious. One for cooking, one for fresh eating, and one for staring at from afar. But Golden Delicious's best days are probably behind it. It's still the apple of choice for baby food (where the clientele doesn't seem to be complaining about the low-acid, middle-of-the-road flavor), and a ripe one straight off an Appalachian tree can still charm, but if it's sweet apples you like, there are more aromatic options.

Granite Beauty

Aliases Dorcas Apple, Grandmother Apple, Clothesyard Apple **Origin** Weare, New Hampshire, 1815. **Appearance** A big, waxy, deep red apple (striped orange and carmine away from the sun) with a peened surface, Granite Beauty is round but noticeably truncated on the top and bottom. It's a bit light for its size, not dense. **Flavor** Sweet and barely tart, with a bizarre curry flavor and an almost metallic finish. **Texture** The slippery slices have a viscous quality. The flesh is tender, but not mushy. **Season** Best in early October; by November, the thrill is gone. **Use** Eat fresh in October. Excellent dried, soft and smooth and nutty. **Region** New Hampshire; exceedingly rare.

Granite Beauty is like the Charles Bukowski of the apple world. It gives the feeling of a dissolute existence brought on by life too deeply felt. The network of pale scarring across the surface, as if you were viewing the Badlands from a plane; the strangely oily skin; the air of noble ruin; Mickey Rourke will play it in the film adaptation.

Part of the reason Granite Beauty looks like it just came off a bar fight is the "peening"—shallow depressions all over its surface, as if it were a metal sculpture shaped by a ball-peen hammer. (Granny Smith sometimes shares this effect.) Everything about the apple is unusual. Many people who taste a Granite Beauty notice its mysterious spice. New Hampshire resident Ben Watson, author of *Cider, Hard and Sweet*, characterizes it as cardamom and curry. I think of it as mellow and cheesy—like the metallic finish of cottage cheese. This sounds pretty unattractive, yet the apple has a certain magnetism.

Although the state of New Hampshire claims the Granite Beauty as its own, Mainers could make a case. The story goes like this. Around 1815, a woman named Dorcas Neull, who lived on a Weare farmstead, visited friends in Kittery, Maine. Traveling by horseback, and finding that she needed a riding crop for the return journey, she reached down and plucked a little apple seedling from the roadside. When she arrived home, she replanted her trusty riding crop and carefully nurtured it. According to Zephaniah Breed, the Weare, New Hampshire, farmer who introduced the apple to the greater world in 1860, "When it produced its first fruit it was found to be excellent, and Dorcas claimed it as her tree. When nephews and nieces grew up around her, the apple was called the Aunt Dorcas apple, from the claim she had upon it. As she grew older and the grandchildren grew up, it took the name of the Grandmother apple. In another part of the town it was called the Clothesyard apple. Believing it to be mostly of a distinct name, we call it the Granite Beauty. The old tree has long been gone, but young trees are plenty in the vicinity."

The Granite Beauty became a minor hit in New Hampshire, and in 1905 Beach could write, "In some portions of New England it is still much esteemed," but it was always a slow-growing variety. Further, as Ben Watson has noted, caterpillars tend to preferentially denude it of leaves. By the twenty-first century, it was virtually extinct, with Gould Hill Farm in Contoocook, New Hampshire, possessing the last four trees in commercial production. So Watson, who calls the Granite Beauty "the apple that only I love," got the Granite Beauty boarded on the Slow Food Ark of Taste and has distributed cuttings to a number of commercial and preservation orchards. With luck, it will no longer be the apple that only Ben loves.

Gray Pearmain

Origin Skowhegan, Maine, 1800s. **Appearance** The flesh inside is ghostly white. The skin glows tennis-ball yellow-green. A basket of Gray Pearmains beckons like a neon bar sign. A netting of tan russet (which is about as close as this apple gets to "gray") usually covers about half the apple, while the sun-exposed side blooms a brilliant fuchsia. Midsize, it is heavy and dense. **Flavor** Lots of sugar, limey acid, and evergreen aromatics, like a pear wrapped in spruce bark. Beautifully balanced. **Texture** Truly crisp and crunchy—like a raw Jerusalem artichoke, not the foam of modern apples. Super fine-grained, which may be the key to its pleasure. Has some of the only russet skin that's truly fun to eat. **Season** Pick in October, eat fresh by Christmas. **Use** Fresh eating, salad, crisps (especially if picked early). **Region** Maine only.

A great apple, and Exhibit A in the case for preserving regional diversity. Almost no one has ever heard of Gray Pearmain, yet it routinely wins blind tastings when it gets the chance. I'd certainly never heard of it. Then I saw a crate of Gray Pearmains at Great Maine Apple Day. They looked pretty and lemonlike. I bought three. Even as I raised one toward my mouth, it was already giving off vibes of deliciousness. The texture, the weight, the coloration—and maybe something more metaphysical. I bit into one and got a mouthful of pear, lemon, and blood orange juice. Wow. Where had it been all my life?

In Fairfield, Maine, as it turned out—and only in Fairfield, Maine. In 1974, Steve Meyerhans bought an old farm from an old fellow named Royal Wentworth. Wentworth, originally from Skowhegan, identified six of the trees as Gray Pearmains, which Meyerhans didn't know. The trees were about fifty years old, and he never learned more of the variety's history, but he noticed its excellence immediately. He grafted new trees, and, though five of the six original trees have now died, the Gray Pearmain lives on solely in Meyerhans's new trees at the Apple Farm, and in John Bunker's grafts that he sells in the Fedco Trees catalog.

Honeycrisp

Origin St. Paul, Minnesota, 1991. (Seedling of Keepsake.) **Appearance** A large, green apple half covered in brick-red stripes. Surprisingly homely for such a rock star. Feels a bit like an old-fashioned Christmas ornament. **Flavor** Sweet and dilute, with a hint of watermelon. **Texture** Seemingly designed by a team of lab technicians and focus groups, Honeycrisp doesn't crunch like normal crisp apples; it shatters in your mouth like an apple-flavored Cheeto, and juice explodes from the bursting cells. The effect is exhilarating. **Season** September to October. Stores well. **Use** Famous for devouring out of hand. A mess in pies and tarts. On the other hand, the flavor intensifies into a fine dried apple. (But if you are drying your pricy Honeycrisps, something has gone wrong.) **Region** Abundant in Minnesota and the Midwest, but Honeycrisps are now found in supermarkets nationwide.

When you bite into a Honeycrisp, one perfect, bite-size chunk cleaves effortlessly into your mouth with a snap. Sweet juice sprays across your taste buds—an effect of the exceptionally large and turgid cells, which pop like caviar. It's the kind of impact product that usually requires the combined efforts of Frito Lay's entire R&D department working overtime for a year. Yet the Honeycrisp is the modest work of the University of Minnesota's apple breeding program, and even they were just along for the ride. When the U of M released the Honeycrisp in 1991, they said it was the offspring of Honeygold, their star apple from an earlier generation, and Macoun, a renowned child of McIntosh. Since the Honeygold was descended from Golden Delicious, this made the Honeycrisp the reunification of two great families of American apples, the McIntosh line and the Golden Delicious line. There was a certain narrative appeal to this story, but it turned out to be completely false. Something was amiss in the records, and DNA testing in 2005 revealed that the Honeycrisp was a cuckoo that had slipped into the wrong nest. Keepsake, another U of M creation, was its real mommy. Its father flew the coop long ago and will never be known, though rumors abound that he came from a line of, gasp, *crab apples*.

Whatever its lineage, the Honeycrisp is a different beast altogether. It truly is, as its marketing campaigns never fail to mention, "explosively crisp." The first time the Honeycrisp's apple breeder tasted it, he says, "the texture was so unusual, I wasn't sure if it was good or bad." Good, everyone has now decided—and it is the best proof I know that texture trumps flavor. A great apple should excel in both, of course, but no flavor, no matter how spectacular, can compensate for a mushy texture. Conversely, the Honeycrisp has a simple flavor; there is no acid flourish, no exotic aromas, just glorified sugar water. But with such explosive pyrotechnics at the front of the stage, the flavor doesn't need to be more than a drum kit, laying down a steady beat of sweetness. I once set a bowl of Honeycrisps and a bowl of Cox's Orange Pippins before my three teenage nieces, intending to demonstrate the superior flavor of the old apples. Instead, the bowl of Honeycrisps evaporated by the end of the day, while the Coxes sat untouched. If you want your kids to eat fresh apples, buy them Honeycrisps (unless, of course, you have rare access to Pixie Crunch).

Other than apple snobs, who whine about the Honeycrisp the way jazz lovers grumble about Kenny G, everybody seems to adore the Honeycrisp. Growers can't plant it fast enough; they sometimes refer to it as the Moneycrisp. In just a few years, it has become the ninth most popular apple in the United States, and it will surely go higher. The University of Minnesota—which collected about a dollar in royalties for every Honeycrisp tree sold by a nursery until 2008, when its patent expired—made ten million bucks off the apple. Although the university is careful to praise the Honeycrisp's "outstanding flavor" in official materials, it has tacitly acknowledged its most famous apple's shortcomings in its promotion of the SweeTango, its newest apple, which could have been named Honeycrisp 2.0—same awesome texture, much better flavor (if you believe the hype).

Hubbardston Nonesuch

Aliases Hubbardston, American Blush **Origin** Hubbardston, Massachusetts, early 1800s. By 1832 it was already a favorite in the Boston region. **Appearance** Regal. Esopus Spitzenberg's taller brother, with natty white stars covering a plush red coat. Sometimes a fair amount of pink or greenish background shows through. Large, ribbed, conic, classic. **Flavor** Quite sweet and aromatic, like bubble gum with a saving trickle of acidity. The raspy skin hints of cucumber. **Texture** Hard and crunchy, like a pear straight off the tree, with unyielding skin. After about a month, its flesh takes on a Mac-like tenderness. **Season** Ripe in September to October. Best by December. **Use** Eat fresh. Makes good cider. **Region** Its best remaining refuges are scattered across New England.

Most apples fall into two categories: the soft early-fall varieties, which begin to go downhill almost as soon as you pick them, and the rock-hard keepers, which are picked in late October or November but need several months of storage before those tight starches relax into sugars. Part of Hubbardston Nonesuch's claim to fame was that it shot the gap; its flavor peaked in October, and by the time it passed its prime around Christmas, the winter apples were ready. Although it never achieved great commercial success, it always had a strong reputation and was an easy sell in nineteenth-century New England.

People have raved about this apple since its early days ("nonesuch" is the English equivalent of the French "nonpareil"—unrivaled) and they continue to do so, but it is a notorious chameleon. In different soils and climes it can become very different apples. I've seen Hubbardstons that were as pink as a Pink Lady, and others that were as mottled and warty as something that crawled out of the Old Testament. The flavor can be sour and herbaceous, with celery-fennel notes, or full-on plum tart, or weak and watery. It has disappointed me more than once.

Hubbardston Nonesuch has a touch of proto–Red Delicious to it, with that pentagonal shape, leathery hide, and Bazooka Joe flavor. Its texture and flavor are superior, but I sometimes think of Hubbardston as the primeval ideal from which Red Delicious fell, and I think the world was a delightful place when apples were named things like Hubbardston Nonesuch.

Hudson's Golden Gem

Origin Tangent, Oregon, 1931. **Appearance** You will know it from other russets by its exaggerated conical shape. Brown russeting covers a light green background, speckled by a constellation of white lenticels. The flesh is quite pale. **Flavor** Very sweet, crunchy, and fragrant, Hudson's Golden Gem eats like an Asian pear. It has almost no acid, yet still manages to be refreshing. **Texture** Crunchy and grainy when young, becoming quite tender once dead-on ripe. **Season** October. Will keep for a couple of months, but it doesn't improve in flavor. **Use** Great for eating fresh or in crisps, where its warm, nutty flavors are enhanced by a squeeze of lemon. Dried, it takes on a date flavor. **Region** First popular in Oregon and Washington, but now a favorite of small orchards across the country.

This apple thinks it's a pear. The shape, the color, the russeting, the intense aromatics. Even the granular texture is pearlike. In fact, after being discovered in an Oregon fencerow in the 1930s, it was originally sold as a pear, and it sort of begs the question as to where you draw the line.

Or, you can not worry about such things and just enjoy it, uncategorized, as an uncommonly delicious thing. Really, this is one of the greats. Easy to grow, a generous bearer, and its apples still come out in decent shape when neglected. It can grow virtually anywhere, frigid cold to downright toasty, but in wetter climates it tends to be fully covered in brown russeting, while in arid areas it has smooth, thin, pale yellow skin.

James Grieve

Origin Edinburgh, Scotland, 1890. Appearance A medium, roundish apple with shiny yellow skin streaked with carmine. When ripe, it flushes one of the prettiest shades of bright red you will ever see. There is usually some runelike russeting around the stem. Flavor Quite tart at first, but this soon fades into a blancmange sort of creamy, sweet, comfort food. Not challenging or tart. The waxy skin leaves a slightly bitter finish. Texture Soft, creamy, juicy. In the words of Bunyard, "melting, almost marrowy flesh." Season September. Eat immediately if you like apples tart, or wait a few weeks for the custardy version. Use Eat fresh, or make a smooth and creamy sauce. Region Mostly seen in the United Kingdom. Very rare in the United States.

This very normal looking apple hides an abnormal eating experience. After a few weeks of relaxing off the tree, it develops some of the smoothest flesh you'll ever find, making it almost like eating a firm custard or, as Bunyard put it so well, marrow. This can be disconcerting if you don't know what to expect, but is actually very rich and pleasant once you make the mental adjustment. James Grieve survives primarily as an excellent pollinator of Cox's Orange Pippin. (Apple trees, being grafted clones, cannot pollinate their own variety—one of nature's ways of avoiding inbreeding—and have varying success at impregnating others.)

Karmijn de Sonnaville

Aliases Karmijn, Karmine **Origin** Netherlands, 1949. (Seedling of Cox's Orange Pippin.) **Appearance** Midsize Karmijn has the same antiquey autumn-leaf patina as its parent, Cox, but with more brick red in the mix, and a little russeting on both the stem and flower ends. **Flavor** If you like citrus, Karmijn is the apple for you. Karmijn takes the taste profile of its parent and turns up the acid to 11, skewing things strongly in the direction of lime and kumquat. It's like a spherical margarita. The skin has some nice astringency, leaving your lips rough and chalky. **Texture** Crisp and crunchy, just like Cox, with coarse, juicy flesh. **Season** September for pain; October for pleasure. Chilled, it stores well all winter. **Use** A howlingly good dessert apple. All that acidity also makes for some of the zingiest fresh cider you'll ever taste. If the acidity is too much for you, leave it alone for a month and it will mellow. Makes a killer crisp. **Region** Happiest in the northern United States.

Apple geeks swoon at the mention of "Karmine," a seldom seen but universally adored variety that is the all-time favorite of quite a few orchardists. It was bred by the apple breeding program at the Wageningen University in the Netherlands in 1949, and released commercially in 1971. Some people thought it would be the next big thing, but it never took off, perhaps because its weathered looks don't appeal to mainstream consumers, or because it struggled in the damp continental climate. It found its true home in the northern tier of U.S. states, where it does well, yet even there it remains a rarity. Why? How could something this good be so obscure? I have no answer, except that names are more important than you might think. Re-released as Hurt So Good, it would be unstoppable.

King of the Pippins

Alias Reine des Reinettes—maybe **Origin** Brompton, England, circa 1800. **Appearance** A compact apple infused with all the colors of a vacation sunset: mango, orange, lightest red. The stem is russeted and the skin is smooth and slightly waxy. **Flavor** There is mango in the flavor, too, along with wine grape, Earl Grey tea, and marzipan. As the weeks of storage go by, those flavors fade and the lychee notes come on strong. The skin provides a welcome bitter lemon-peel finish. **Texture** Although never extremely crisp, it is juicy and pleasing and holds its own through October, beginning to soften after that. **Season** Early September (late August if you want it tart and crunchy). Eat fresh. Done by Thanksgiving. **Use** One of the world's great dessert fruits. **Region** Still a favorite in England, yet strangely rare in the United States.

In 1881, the British apple industry was in crisis. Britain was a land of small-scale orchards, each growing a hodgepodge of varieties, many going by strictly regional names. Most tended to be the small, often russeted, dull yellow and green apples that did well in the British climate. And they were getting destroyed in the marketplace by the onslaught of imported American apples—large, bright red Baldwins from New York, Rome Beauties from Ohio, and Ben Davises from the South, along with bright green Newtown Pippins from Virginia. As always, looks triumphed over inner virtue, and so a national campaign was begun by the British Pomological Society so that, as one prominent nurseryman put it, "the public will soon learn to discriminate between the brightly coloured but dry and flavourless American kinds and fresh home grown apples." This culminated in 1883 in the Royal Horticultural Society's National Apple Congress in London, at which an astounding 1,545 British varieties were displayed. As part of the festivities, favorites were chosen. The winner? Cox's Orange Pippin, one might guess? That was number two, just ahead of its parent, Ribston Pippin. Blenheim Orange? Number five. The most popular apple in Britain was King of the Pippins.

With King of the Pippins you are indeed in the upper echelon of flavor. Back in 1851, Dr. Robert Hogg, England's leading apple expert, called it "one of the richest flavored early dessert apples, and unequalled by any other variety of the same season." It packs incredible complexity into a small package. You virtually never see King of the Pippins in the United States, because it almost always gets called Reine des Reinettes here. Most authorities believe they are one and the same apple. I'm not so sure. The ones I've seen, from the USDA orchard in Geneva, have a stunning orange glow to them, significantly different from the milk-paint red of the Reine des Reinettes. (Further confusion: King of the Pippins in England look nothing like what the USDA has in its database.) The flavor, however, is pretty close; both have loads of sweetness, acidity, and tropical aromatics. These are two of the best apples on the planet, and you are lucky indeed if you ever get the opportunity to compare them side by side.

Lamb Abbey Pearmain

Origin Lamb Abbey, Kent, England, 1804. (Seedling of Newtown Pippin.) **Appearance** Lime-size, cute, round, splattered with red-orange paint and white lenticels, like a baby Gala. **Flavor** Very sweet, tart, zesty, tropical, crunchy, with a touch of pineapple-banana bubble gum. **Texture** Medium crisp and very juicy. **Season** September. **Use** Dessert fruit. **Region** Famous but rare in Britain and New England.

Lamb Abbey Pearmain is flat-out yummy. It has the Juicy Fruit jelly bean flavors of Fuji and Gala, with more zing and better crunch. Zeke Goodband, the apple guru at Vermont's Scott Farm, calls it, "One of the most delightful apples in the world." Its small size has always made it unattractive to commercial growers, and in fact it had disappeared from English culture until the English epicure Morton Shand rediscovered it in the 1940s as part of his patriotic campaign to revive English heirloom apples. Interestingly, the original seed—which was planted by Mary Anne Malcolm in her garden in Lamb Abbey, Kent, in 1804—was from a Newtown Pippin that had been imported from America, making Lamb Abbey Pearmain one of the first English varieties with an American heritage. Its pineappley flavors were an immediate hit, and in 1819 Mary Anne Malcolm received a medal for her work from the Horticultural Society of London.

Macoun

Origin Geneva, New York, 1923. (Cross of McIntosh and Jersey Black.) **Appearance** The richest burgundy. A grove of ripe Macouns, covered in purple bloom, looks royal and luxurious. The Mac-size fruit has five subtly visible ribs. The flesh is pure white. **Flavor** Very sweet, very tart, with an uncomplicated yet addictive lemon zing that makes your mouth tingle. This is the Sprite of apples. **Texture** Seems to shatter into juice at first pressure, and you are left with almost nothing to chew. It just disappears, like cotton candy. This is lovely and strange. If not perfectly stored, it'll get soft fast, so think of Macoun as a seasonal treat. **Season** September to October. **Use** Eat it standing under the tree. Swoon. **Region** A cult favorite in the Northeast.

Another naming debacle from the branding-challenged folks at Cornell. People rave about this apple, even when they can't quite pronounce it. (Not muh-COON. Some say muh-COW-in, some muh-COWN; its eponymous horticulturalist, W. T. Macoun, was Canadian, which should give you a clue, eh?) This great apple—the best thing to ever spring from a McIntosh seed—might have conquered the nation ninety years before Honeycrisp if only it had been named something like Purple Haze. Instead, it has flown under the radar, filling the baskets of apple insiders while struggling in the market. Because you can't take it with you—Macoun neither ships nor stores—it functions like a seasonal madeleine in the memory of many a New Englander.

Mother

Aliases American Mother, Gardener's Apple **Origin** Bolton, Massachusetts, 1840s. **Appearance** One of the few apples to truly achieve the coloration of a mango, a real yellow-orange, overlaid in shiny red. Mother is medium-size, sometimes round and sometimes quite conic. **Flavor** Sweet and complex, with a bit of pine on the finish, and very little acid. Like Esopus Spitzenberg's sweeter, gentler sister. **Texture** Tender and creamy in September (South) and October (North), it goes soft by November. **Season** September to October. The lack of acid means it does not keep long. **Use** Eat fresh by November at the latest. Too soft for cooking or keeping. Unshippable. **Region** Forgotten backyards of America. Rabun County, Georgia, used to be a Mother hotspot.

Mother is one of those yummy apples that has been laboring in anonymity for more than a century. It never achieved much popularity in its native New England, where the tree tends to struggle, but it became a minor hit in the South, especially Georgia, and in England, where Edward Bunyard, who was skeptical of American varieties, sang its praises: "Its remarkable flavour, reminiscent of the pear drops of our youth, stands alone, and its mellow flesh at the right moment is greatly to be recommended." Tom Burford lists it among his Top 20 Dessert Apples.

Mother is a classic example of the small-celled "melting" or "marrowlike" texture that was once esteemed in apples, but seems quite alien to our modern palate, which has learned to expect explosions. I'm not sure what was flavoring the pear drops of Bunyard's youth, but most commentators on Mother mention an unusual flavor, comparing it to things like wintergreen and spruce. It's worth finding, though that isn't easy; the apples are too soft to wander far from the tree, so Mother remains a nice discovery for the backyard orchardist—and for friends of the backyard orchardist.

Mutsu

Aliases Crispin, Crispen **Origin** Aomori Apple Experiment Station, Japan, 1937. (Cross of Golden Delicious and Indo.) **Appearance** A monolithic green behemoth with butter-yellow flesh and an occasional orange cheek. **Flavor** Very, very sweet, with pear-water aromas reminiscent of Golden Delicious. There is some tartness and even a touch of astringency, but mostly it's like eating a watermelon. **Texture** Ultra-crisp flesh that snaps apart to release runnels of clear fruit juice. The skin is surprisingly chewy. **Season** October. **Use** Famously good fresh, but equally excellent in pies. Makes a phenomenal (though hard to finish) baked apple. **Region** A standard in supermarkets from Topeka to Tokyo.

Like the Incredible Hulk, Mutsu is huge, green, and strangely lovable. That massive bulk hides a sweet demeanor. You wouldn't call it complex, but Mutsu is reliably great. Anywhere you see it, you are guaranteed a joyous crunch fest filled with Golden Delicious–style honey aromas. In Tokyo, you'll see freakishly perfect Mutsus that were grown with paper bags around them to protect from any injury or insect, packaged in gorgeous boxes and selling for the price of a fine kitchen knife. The apple was named for the old province of northern Japan where the Aomori research station lies—Japan's prime apple country.

Nodhead

Aliases Jewett's Red; Jewett's Fine Red **Origin** Hollis, New Hampshire, circa 1780, property of Samuel Jewett. **Appearance** The more you look at Nodhead, the more stunning you realize it is. Striped yellow and orange against a pale background on one side, it completely transforms on the other, with dark red skin and a purple bloom worthy of a Macoun. It has huge white lenticels, as big as any I've seen. Slightly flat and midsize, it has five distinct sides, like a pentagon. **Flavor** Sweet, with just a hint of tartness, and lovely, buttery aromas of melon and pound cake. **Texture** Old-school crunchy, like a sweet water chestnut. It snaps off in chunks that must be demolished with the jaw. Juicy and refreshing, with a very fine grain. The skin is quite thick and hard. **Season** "It is excellent eating right off the tree after a few frosty nights in October," says John Bunker, "and it remains crisp and tart well into winter." **Use** Eat fresh, good in pies and crisps, store for later. **Region** Maine and New Hampshire hold most of the remaining examples of this very rare apple.

There was a time when crisp was not so valued in an apple, and in that time, Nodhead was held in great esteem. It's a great example of an old-timey apple, and an insight into old-timey palates. Modern apples like Honeycrisp and Gala tend to have large, juice-filled cells that shatter when you bite them, spraying your mouth with sweet liquid and virtually dissolving. The older preference was for apples with fine-grained cells packed tightly together. Dense and substantial, they required a lot more work from the eater. Nodhead is one of those. The apple resembles Blue Pearmain in both color and scent and is rumored to have been a seedling of it. Popular in certain towns of Maine and southern New Hampshire in the 1800s, and possibly the 1700s, the trees were never productive enough for the Nodhead to go national. Still, this must be one of the older American apples, because Samuel Jewett died in 1791. (One Hollis historian puts its origin at around 1780.) He was a Revolutionary War vet, having fought at Bunker Hill and elsewhere, and he may have suffered post-traumatic stress disorder, or maybe he just had Parkinson's disease; the people of Hollis nicknamed him Nodhead for his habit of constantly bouncing his head when he walked or talked.

Orleans Reinette

Aliases Reinette d'Orléans, Winter Ribston **Origin** France, 1776. **Appearance** Wide and squat, like a tire. Dull green when young, with russeted brown dots scattered over the surface. As fall progresses, it turns old-barn red on the sunny side, wan yellow beneath. There is usually an orangish brown russet around the stem. **Flavor** Rum punch, heavy on the lime and nutmeg. **Texture** Its coarse white flesh is hard and crunchy, just juicy enough to get the job done. The skin, quite chewy and persistent, sticks around interminably. Peel it. **Season** Ripe in October. By Christmas the deliciousness has waned. **Use** Savor it fresh. You could do worse than follow the lead of Edward Bunyard, the British nurseryman: "As a background for an old port it stands solitary and unapproachable." Excellent dried. Great for cooking, where its zesty quality makes it a good foil for meat. Try it with roast pork, or grated into a salsa with pork tacos (see page 274). **Region** Popular among apple enthusiasts in both Europe and North America.

This apple's scaly coat and wizened exterior is either profoundly unattractive or full of character and wisdom, depending on your bent. For Bunyard, "Its brown-red flush and glowing gold do very easily suggest that if Rembrandt had painted a fruit piece he would have chosen this apple."

"Oranges and walnuts" is the traditional description of Orleans Reinette's big flavor, and that's certainly in the ballpark, but to me that misses the *guarapo* grassiness of its youth. That limey zing, combined with the dark rum spice and the nutmeg, brings to mind Cuba Libres (nutmeg is one of the major flavor notes in Coca-Cola). All agree that the flavor is wonderfully complex, though for some the touch of dryness drops it out of the top tier of apples. Not for Edward Bunyard, who placed Orleans Reinette atop the entire heap. "This stands of all apples highest in my esteem," he wrote in 1929 in *The Anatomy of Dessert*. In this he joined the nation of France, which has long regarded Orleans Reinette as the crème de la crème of *pommes*. I hate to pile on, but I'll add that of the dried apples I've tried, Orleans Reinette was the best, very dry and chewy at first, then, as it loosened up, wonderfully tart and fruity. Think of it as the natural chewing gum of French children, circa 1800.

Ozark Gold

Origin Springfield, Missouri, 1970. (Golden Delicious seedling.) **Appearance** Like a Golden Delicious with makeup. Ozark Gold's blemish-free skin is more golden than its mother's, and it usually has a perfect rosé blush on one cheek. This large apple has a modish vase shape. **Flavor** Ozark Gold's juicy, yellow flesh has the same gentle honey flavors and wafting perfume as its parent, but then, at the end, a surprise: a rich marzipan finish lingers on the tongue. **Texture** Tender, but not mushy. The skin is a tad tough. **Season** Late August in the South, September in the North. **Use** Eat fresh. Too mild for good sauce or cider. **Region** Farmstands in the southern Midwest and Appalachian Mountains.

There are a number of apples that were born in the North but do better in the South (such as Newtown Pippin and Winesap), but this charmer is one of the few that bucks the trend. Although it was developed by the Missouri State Fruit Experiment Station to be a late-summer apple for warm regions, it stays crisper, has more balanced flavor, and is more likely to produce its winning blush when exposed to the chilly nights of the north (or the Appalachian Mountains). Too often in the South, it is soft and weak flavored.

The Missouri State Fruit Experiment Station calls Ozark Gold its greatest success among apples, which is faint praise. The station was tasked "to combine the indestructible characteristics of 'Ben Davis' with the higher quality and choice flavors of 'Red Delicious' and 'Jonathan,'" because in 1904 the state of Missouri had 25 million apple trees, the most of any state, and most of them were Ben Davis, which was fast cementing its reputation as World's Worst Apple. Ozark Gold is a fine apple for roadside stands, but nothing the station produced was able to change the fate of Missouri apple farmers.

Pink Lady

Alias Cripps Pink **Origin** Western Australia, 1973. (Cross of Golden Delicious and Lady Williams.) **Appearance** A long, tall drink of water, with uniform reddish pink skin and thousands of white freckles barely visible beneath the tan. **Flavor** Sweet, tropical, and nicely tart. **Texture** The queen of juice. Very crisp. Pink Lady doesn't quite dissolve like Honeycrisp and the other modern apples; it's more substantial, and the skin is notably thicker. **Season** The Queen of Pink knows no season; a well-manicured Pink Lady from one hemisphere or the other is always waiting for you in the supermarket crisper. **Use** Good for fresh eating, excellent in salads (resists browning), and great for culinary purposes (no shrinkage). **Region** Requires hot, sunny climates to develop its characteristic blush. Big in Australia, South Africa, and the western United States. Sold worldwide.

Pink Lady has the dubious honor of being the world's first "club apple," meaning not just the trees, but the individual fruit is tightly controlled by the Australian apple consortium that released it. Every stage of its growth and sales is regulated. This was done so as not to dilute the brand: Pink Ladies are tested for color and sugar/acid balance, and everything that doesn't make the cut (more than half the production, in fact) is sold as generic Cripps Pink, which is the name of the variety. That guarantees (in theory, anyway) that every Pink Lady you buy is going to be high quality, and you pay a premium for that. Yet somehow half the Pink Ladies in the marketplace have all the pizzazz of a bottle of apple juice. Still, there's no holding back the Pink tide; since being released in the United States in the late 1990s, it has ripped past such mainstays as Cortland, Braeburn, and Jonathan to become a top ten apple. Every now and then, you get a truly ripe Pink Lady, and you know it: a crunchy mango bursts in your mouth, and the angels start to sing.

Pinova

Aliases Piñata, Sonata **Origin** Dresden, Germany, 1986. **Appearance** A very smart-looking apple with smoothly rounded shoulders tapering to a tight foot. Red stripes and blush cover a creamy yellow background. **Flavor** Very much in the tropical Gala camp. Very sweet. So much is going on that you don't notice the tartness, but it's there, keeping things refreshing. There is a little banana and coconut, and a lot of carrot juice. **Texture** Quite crisp. Lovely, bright juice oozes from the cells. The skin is a bit crunchy. **Season** Fall, but available nearly year-round through the magic of controlled atmosphere storage. **Use** Superb out-of-hand, but also good for all culinary purposes. Won't brown in salads. **Region** Grown almost exclusively in eastern Washington, sold everywhere.

Of all the nouveau apples that have transformed the supermarket apple section in recent years (Fuji, Gala, Braeburn, Pink Lady, Ambrosia, Jazz, Honeycrisp, etc.), this is the best. Pinova has Golden Delicious, Cox's Orange Pippin, and Duchess of Oldenburg in its lineage, and all three factor into its spicy, tropical, honeyed, snappy, striped identity. Somehow, it has both the modern, fusion-juice appeal of the new designer apples and the high citrusy notes of the old English classics. The apple was bred in Germany as Pinova, and in the 1990s a few early adopters in the United States got their hands on the apple, which has struggled with identity issues. At first, it was sold in the United States as Corail. That proved a nonstarter, so the name was changed to Sonata. Many growers preferred to stick with the German Pinova. In 2004 Stemilt Growers, a major apple producer in Wenatchee, Washington (apple capital of the universe), bought the exclusive rights to the apple and renamed it Piñata—a combination of Pinova and Sonata, see? Growers that already had the apple were allowed to keep selling it as Pinova. None of these names are winners (Do I really want to imagine whacking my apples with a stick as they hang from the tree? Maybe flavor is supposed to burst out?), but whatever you call it, it's a great apple.

Pitmaston Pineapple

Origin Pitmaston, near Worcester, England, early 1800s. **Appearance** A smoothly conical, little golf ball of an apple. Its russet skin makes dry-leaf noises when you rub it. When fully ripe, it turns golden khaki. **Flavor** There's pineapple for sure, and pine needle, too. "Tastes like Christmas," according to a friend of mine. A smoky, raisiny finish seals the deal. **Texture** A wonderfully crunchy, snack-size package. **Season** September. **Use** Eat fresh. Too small and delicious to bother with any other application. **Region** Grown by a handful of hobbyists in the United States and United Kingdom.

Apple names can be red herrings. The Winter Banana assuredly does not taste like banana. Does the Chenango Strawberry have a hint of its eponym? Hmmm, maybe. But the Pitmaston Pineapple is absolutely appropriate. It should feature in luaus, and upside-down cake. This diminutive apple was introduced to the world at the 1845 Exhibition of the Royal Horticultural Society by a Mr. Williams, who had a nursery in Pitmaston, but he explained that it had been raised from a seedling sixty years earlier by the steward to Lord Foley, one Mr. White. Perhaps due to its diminutive size, it never caught on in England and was almost gone by the time Bunyard called it "a remarkable blend of honey and musk" in *The Anatomy of Dessert* and helped save it from oblivion. In 1920, he wrote, "This is one of the old fruits which have been neglected on account of their small size, but its distinct and delicious flavour should give it a place in the gardens of connoisseurs." And so it should. And does. The Virginia apple connoisseur Tom Burford lists it among his Top 20 Dessert Apples. It makes my top five.

Pixie Crunch

Origin Lafayette, Indiana, 1978. **Appearance** Small, round, red, cute. The flesh is nearly amber with flavor compounds. **Flavor** Intensely sweet and strangely savory, as if it has an ounce of caramel in it. It's almost like eating a candied apple. **Texture** Honeycrisp, eat your heart out. Nice thin skin, and lots of juice. **Season** September. Doesn't keep. **Use** Eat fresh; smile. **Region** Strangely rare, yet well adapted to the Midwest and other temperate climates.

If I was designing a fruit to beguile every child in the world, I would design the Pixie Crunch. Crisp as a chip, sweet as candy. Fits perfectly in a pint-size hand. Irresistibly red. Then I would come up with an irresistibly cute name like, well, Pixie Crunch. But I wouldn't bother doing any of this, because the PRI (Purdue/Rutgers/Indiana University) Apple Breeding Program already did it, channeling Red Rome, Melba, Golden Delicious, Edgewood, Rome Beauty, Crandall, and a crab apple to get it. That was forty years ago, yet PRI didn't even bother naming and releasing the apple until 2004. Why isn't there a Pixie Crunch in the lunch box of every schoolchild in America? Commercial growers think it's too small. They obviously haven't asked America's kids, who find most apples too big. Want to turn kids off apples for life? Hand them a huge, thick-skinned Red Delicious. Want to turn them into apple-obsessed fruit bats? Drop a Pixie Crunch into a pixie palm. "To the young," Edward Bunyard wrote in 1929, "the crunch is the thing, a certain joy in crashing through living tissue, a memory of Neanderthal days."

Pixie Crunch is ideal for your backyard. It's scab-resistant, stays small, and bears heavily. The tree has a Dr. Seuss vibe. Your kids can handle all the picking duties. They can't graft it, however, and neither can you; like many new apple varieties, Pixie Crunch is patented.

Pomme Grise

Aliases Pomme Gris, Gray Apple, Pomme Canada, Reinette Grise **Origin** Montreal vicinity, St. Lawrence River Valley, Quebec, 1800s. **Appearance** Small, but heavy for its size, like a golf ball. The tan-green russeting covers the entire fruit. **Flavor** A classic russet, tart and grassy when young, maturing to a wonderfully deep, nutty, marzipan richness. **Texture** Dense and crunchy. Not especially juicy. **Season** Pick in October. **Use** Excellent fresh. When grown in the neighborhood of the 45th parallel, it makes a fine keeper, but not in locations farther south. **Region** Rarely seen outside of Quebec and northern New England.

The Gray Apple is a famous old Montreal variety grown throughout the St. Lawrence River Valley in the 1800s. No one is certain whether the apple started as a wild seedling in Quebec or was a Reinette Grise (a French variety dating to the 1600s) brought by settlers from France or Switzerland. I suspect it was a seedling, because the Pomme Grise is so happy in frigid Quebec winters. It popped over to the U.S. side of the St. Lawrence, too, but never develops its best flavor much south of there. In the right spot, however, it is quite delicious. The celebrated landscape designer and horticulturalist A. J. Downing called it "undoubtedly one of the finest dessert apples for a northern climate," in *The Fruits and Fruit Trees of America* (1845). S. A. Beach (1905) also praised Pomme Grise in *The Apples of New York*, but pointed out that there was no market for a shrimpy gray apple, no matter how tasty.

Reine des Reinettes

Alias King of the Pippins—maybe **Origin** Netherlands, late 1700s. **Appearance** A gorgeous antique apple with striped salmon and orange and carmine skin, about half covered in the finest tan russet. The russeting is strongest at the stem and flower ends. The size and hardness makes it feel like a baseball in your hand. The skin is noticeably dry and parchmentlike. **Flavor** Exploding with lychee and lime. Lots of grapefruit and other citrus. **Texture** A crunch fest. This is an extremely hard apple, yet it crumbles satisfyingly into small bits. The toothy skin has a little resistance to it. **Season** October. **Use** Top-notch dessert apple. Also great for pies and cider. **Region** A favorite in England in the 1800s. Still a must-have for heirloom orchards in the United States and United Kingdom.

Some apples have an ineffable quality that pomologists refer to as "lively." To me, it means that the flavor engages with the eater, comes alive in the mouth, and startles the taste buds. It demands to be thought about. Reine des Reinettes is lively. Once an important Norman cider apple, probably brought over from the Low Countries, this Queen of the Reinettes (but not Queen of Princesses; see the glossary for more) holds its own with the best in tastings. Many experts believe it is the same apple as King of the Pippins.

Ribston Pippin

Origin Ribston Hall, Yorkshire, England, 1688. **Appearance** With the same weathered fall-leaf pattern as Cox, Ribston likes to cover itself in a little more red and stays a bit smaller. Russeting fills the stem bowl. **Flavor** Aromatic and citrusy, like Cox, but damped down. Meaner, greener, and more earthy. **Texture** Hard and dry. **Season** Pick in September. Eat within a month. **Use** Good for fresh eating. Makes a bone-dry, lemony, gaunt hard cider. **Region** Still common in the United Kingdom. In America, Ribston can be found south and north, west and east, though never frequently.

One of the most famous of English apples, Ribston had a fine career in the early 1800s before becoming eclipsed by its child, Cox's Orange Pippin. It has suffered a bit of an Archie Manning complex ever since. Back in the day, the "Glory of York" was a big deal. In 1851, England's pomological champion Robert Hogg called it, "An apple so well known as to require neither description nor encomium," adding, "There is no apple which has ever been introduced to this country, or indigenous to it, which is more generally cultivated, more familiarly known, or held in higher popular estimation, than the Ribston Pippin."

The apple began when Sir Henry Goodricke, the lord of Ribston Hall, was touring Europe in 1688 and became smitten with an apple he consumed while in Rouen, Normandy. Back in Yorkshire, he had the seeds of that apple planted on the grounds of Ribston Hall, his sweet spot originally bequeathed to the Knights Templar in 1217. Poor Henry had no grasp of grafting, but to his luck, one of the seedlings turned out to be even better than the parent, though somehow its inherent greatness went overlooked for a century (a good reminder to us all). By the late 1700s, it was much celebrated in London, and by the 1800s it was a leading variety. And it might have stayed that way, were it not for that meddling kid, Cox's Orange Pippin, which delivered all Ribston's roiling aromatics in a sweeter, larger package. Since Cox's ascendency, Ribston has been a bit of a museum piece. It must be sick of being introduced as "Cox's parent." On the other hand, a certain gruff minority, always suspicious of Cox's flamboyancy, maintains that the parent is the better fruit. The English connoisseur Morton Shand, for example, in 1949 recalled his father's strong belief that Cox "lacks the austere aristocratic refinement that Ribston exemplifies transcendentally." I've found nothing transcendent in my limited consumption of Ribston Pippin. Austere, yes. Delicious, no. Other American commentators agree. Maybe it's our Yank uncouthness. Or maybe Ribston just doesn't come out right on this side of the pond. A. J. Downing, editor of *The Horticulturalist*, suspected as much in 1845: "The Ribston Pippin, a Yorkshire apple, stands as high in Great Britain as the Bank of England… But it is scarcely so esteemed here, and must be content to give place, with us, to the Newtown Pippin, the Swaar, the Spitzenburgh, or the Baldwin."

Silken

Origin Summerland, British Columbia, Canada, 1999. (Cross of Honeygold and Sunrise.) **Appearance** Pale green skin fading to white-gold when ripe. Shines like a light in the sun. Sometimes russet fills the stem bowl and drips over one shoulder. **Flavor** Sweet and highly aromatic, with enough tartness to get by. Refreshing but neutral. **Texture** Wonderfully crisp, juicy, and insubstantial. **Season** Early September. **Use** Eat as fresh as possible. Bruises easily. **Region** Although still largely unknown, Silken is gaining popularity in pick-your-own orchards, particularly in the Pacific Northwest, where its early arrival, tender crispness, and China-doll delicacy make it stand out.

This is the closest I know to a *white* apple; the skin gets so pale and translucent that the snow-white flesh shows through. The word almost invariably used to describe Silken is "porcelain," and that's what I would have called it. The fruit has a fragile albino feel to it, and it certainly bruises if handled roughly. Yet it is a delight to eat, a crisp flash of juice and lightness, gone before you know it. This, combined with the apple's brief tenure in earliest September, gives the Siken an ephemeral quality, like a mayfly, and it could have easily fallen into the Summer Apples category. The perfect Silken moment would be to eat one straight off the tree, reflecting for an instant on the end of summer and the first chill of fall. Then the moment's gone, and the apple with it, and by the time you experience it again, another year will have passed you by.

St. Edmund's Russet

Alias St. Edmund's Pippin **Origin** Bury St. Edmunds, Suffolk, England, 1875. **Appearance** A tree full of St. Edmund's Russets is captivating and mysterious. This unusually large russet, round and slightly flattened, is the warm color of a paper bag lit from within by candlelight. Often it is stippled with fine black dots, a natty combination that gives St. Edmund's Russet a formal feel. **Flavor** Like vanilla pudding infused with pear essence. Early in the season, the richness can be masked by a blast of lemony acid, but this gives way to a yellow-cake flavor. **Texture** Crisp and juicy at first, with a fine, creamy grain, becoming airy and light as September wanes. **Season** September. **Use** Best fresh. **Region** Long popular among English apple collectors, its star is now rising in the United States.

St. Edmund's Russet may caucus with the apples, but at heart it's an Asian pear. Its ultra-fine flesh and rich perfume make it one of the great eating apples—just get it early in the season, because it all falls apart fast. The apple comes and goes within a couple of weeks, and there are those (primarily in England) who approach the hunt for St. Edmund's Russets with the zeal that some reserve for chanterelles and truffles. It originated in Bury St. Edmunds, the Suffolk town where rest the remains of King Edmund, slain by the Vikings in A.D. 869. A cult grew up around Saint Edmund, who was believed to deliver the occasional miracle to pilgrims who visited the abbey in Bury St. Edmunds, and one could be forgiven for seeing something sanctified and otherworldly in the inner glow given off by this classy apple.

Sweet Sixteen

Origin St. Paul, Minnesota, 1973. (Cross of Malinda and Northern Spy.) **Appearance** A huge, heavy, tall green apple striped in rose-red. Raised lenticels create a stubbly effect. The flesh is a pale, dilute orange. **Flavor** A misty explosion of melon and bubble gum. Satisfyingly sweet, passingly tart, Sweet Sixteen keeps things simple and fruity, a kindred spirit to the watermelon. **Texture** Crisp, refreshing, and really juicy. This apple gushes. **Season** September. **Use** Eat fresh. Softens if cooked. Keeps well. **Region** Does best in northern climes.

An absolute delight to eat straight off the tree, Sweet Sixteen is one of the triumphs of the university breeding programs. Yet it was never pushed by the University of Minnesota, perhaps because it was never patented, so there was little money to be recouped. Some people say the apple tastes like anise or bourbon. I'm not convinced. John Bunker, who helped popularize Sweet Sixteen after it got left on the back burner, says it tastes like cherry Life Savers, which hits closer to the mark.

SweeTango

Origin St. Paul, Minnesota, 2006. (Cross of Zestar and Honeycrisp.) **Appearance** Like a heavily freckled nectarine in shape and color, with red and peachy swirls over a yellow background. **Flavor** Very, very sweet, with the same generic tropical fusion-juice notes of other modern apples like Gala, and that same weird, aspartame-like aftertaste. **Texture** The platonic ideal of crisp and juicy. An almost religious experience (and a very loud one). The skin is not insubstantial. **Season** September. **Use** Eat fresh. Too pricy for cooking. **Region** Mostly grown in Washington State, Minnesota, and New York, and available nationwide, but still uncommon.

The child of Honeycrisp and Zestar (a can't-miss apple from the 1990s that somehow missed), SweeTango combines extraordinary crispness with mouth-filling fruitiness. The University of Minnesota, which bred it as the successor to the Honeycrisp and expects great things of it, has sunk unprecedented resources into the apple, and has taken some hits for doing so. Like Pink Lady, SweeTango is a "club apple"—it can be cultivated only by a handful of growers who join a consortium called Next Big Thing and agree to strict rules on how the apple can be grown and marketed. The university owns patents on both the trees and the individual apples and receives royalties on every sale. No apple has ever arrived with such fanfare, including a fawning *New Yorker* feature.

Yet thus far, after several years on the market, despite being the only apple I know with its own website and Facebook page, it has failed to re-create the Honeycrisp buzz, and I think I know why. Who on earth thought SweeTango was a catchy name for an apple? Heads should roll. (Instead, SweeTango seems to have set a dubious capitalization precedent: Cornell's apple breeding program, still playing catch-up, recently announced that its next two apples, the sixty-fifth and sixty-sixth in its storied history, will be named SnapDragon and RubyFrost.)

If SweeTango does end up in the dustbins of history, alongside New Coke, *Ishtar*, and Joba Chamberlain, it'll be a shame, because it's a worthy apple. The flavor is too saccharine, too eager to please (I actually prefer Honeycrisp's vapidness), but the boom-crash of it is like listening to Wagner. This is a whole new organoleptic experience, and a consuming one. Perhaps you meant to take only one bite, a sample, but before you know it you've taken another, all those syrupy cells detonating in your cheeks, and then another, juice dribbling down your chin, you just can't stop crunching, and suddenly you are standing there with a skinny core in your hand, looking for a wet wipe.

Tolman Sweet

Alias Talman Sweet **Origin** Dorchester, Massachusetts, late 1700s. **Appearance** Unremarkable. Green, turning yellow in storage, round, medium in size. Often there is a characteristic green "suture line," like an indented, dark green surgical seam, running from the stem to the eye. **Flavor** Sweet and aromatic. Not a trace of acidity. The deliciously strange flavor has elements of Calvados, chanterelle ice cream, and a pear that fantasizes about being a pumpkin. **Texture** Dry, firm, and granular. **Season** October to November. **Use** Eat fresh or bake. In sauce, its skin breaks up beautifully, and it needs no sugar. It should dry nicely, too. **Region** Northeast, especially Maine. A lover of cold.

This dense apple, once quite famous in Maine, and Henry Ward Beecher's favorite for baking, is the antithesis of the modern, explosively crisp apple. In the wrong company, it can be scorned for lacking tartness, juice, or snap, but in the right setting, it can be the most wonderful apple in the world. For me, that setting would be a filthy November night, sleet spitting against the windows. You and a few friends have just finished some rich supper, and you sit around the hearth and pour from a bottle of Calvados or Cognac, and you each eat one perfect, brandied Tolman Sweet as you tell stories late into the night, and the idea of ending the evening and stepping out into the slush starts to seem more and more unfathomable.

Wagener

Origin Penn Yan, New York, 1791. **Appearance** Wagener can be a green ugly ducking when young or a yellow-skinned, scarlet-cheeked hottie when fully ripe. Often you see it still greenish yellow in farmers' markets, and it will be less aromatic. The flesh is snow white. **Flavor** Uncut, it smells a bit like a tomatillo, but there is also a highly unusual spicy aroma, a mango chutney smell, coming from the stem end. The taste is like a sweet and slightly astringent persimmon, less cloying due to some tartness. **Texture** Very juicy and tender, yet sometimes marred by a softness bordering on mealy. **Season** October. **Use** Fresh eating when red, extraordinary pie when green. Makes sweet, aromatic fresh cider. An excellent keeper. **Region** Nationwide in good collections, especially in Upstate New York and New England.

The seed that would give birth to the first Wagener apple tree was taken from an apple in Dutchess County, New York, in 1791 by George Wheeler when he moved to Penn Yan as part of a group of Methodists who believed the Finger Lakes region to be a new Promised Land. There, Wheeler hacked a farm out of the wilderness and planted a seedling orchard on it. The region did indeed prove to be a promised land for the little Wagener seedling, which thrived in the fertile soil and temperate air and soon produced bumper crops of uncommonly delicious red fruit. Abraham Wagener, a neighbor, bought Wheeler's seedling orchard in 1796 and transplanted the best trees to his home in the newly formed Penn Yan village. The apple's fame spread locally and then, when its excellence was recognized by the New York State Agricultural Society in 1847, it went national—and even international: Edward Bunyard was singing its praises in England in 1929. In colder climates, or when not properly thinned, the fruit doesn't develop its red color or full size, yet it still tastes darned good. The tree is a natural semi-dwarf; it bears at an early age, bears a lot, and stays fairly small, all of which makes it a dream for the apple hobbyist. It was just enough of a labor of love to be abandoned by most commercial growers in favor of more foolproof fruit, yet there are still a few Wagener evangelists out there.

Wealthy

Origin Excelsior, Minnesota, 1868. (Crab apple seedling.) **Appearance** Large, round-oblate, with a light red coat over greenish yellow skin. **Flavor** Middle of the road. Good, sweet-tart, fairly one-dimensional, like a McIntosh without the whimsy. **Texture** So crisp, so juicy when young. It has that odd foamy quality, like modern apples; it seems strangely light and insubstantial, and its cells give way without a fight. **Season** October (September in the South). Get it before it softens. **Use** Fresh eating, sauce. **Region** Northernmost states, especially Minnesota.

In 1853, a clever man named Peter Gideon arrived in Excelsior, Minnesota, on the shores of Lake Minnetonka, and planted an apple orchard. To his dismay, he soon discovered that most of the apples popular in the eastern and southern United States were simply not going to cut it in the great white north. Others concurred. Horace Greeley, anti-slavery crusader and publisher of the *New York Tribune*, who had famously advised, "Go west, young man!" modified that sentiment in an 1860 speech in the state, saying, "I would not live in Minnesota, because you cannot grow apples there." Well, Gideon wished to live in Minnesota, and wished to grow apples there, but by 1861 everything he'd tried had failed, and he was down to his last tree (a Siberian Crab) and his last few dollars. He spent those dollars on a bushel of seeds from Bangor, Maine, which he figured had proved their genes in the cold, and he crossed those seedlings with his Siberian Crab, which he knew could survive anything short of a new ice age. The most successful of his crosses was a tough little bugger that he named after his wife, Wealthy Hull. As hardy as Thor, Wealthy went on to become a prized northern apple, and even a popular southern one, and Gideon went on to found the apple breeding program at the University of Minnesota.

(Epilogue: In 1871, Greeley again visited Minnesota, and at the Minneapolis fair was confronted with a table covered in local apples beneath a banner bearing the words of his previous quote.)

In many ways, Wealthy is a quintessential apple. At any moment other than its peak, it embodies the simplistic flavor Edward Bunyard had in mind when he sneered, "For America, in its adolescent stage, schoolboy palates were well satisfied by the Baldwins, Jonathans, Wealthys, and the like," but if you catch it right, it can be sublime, particularly for a boreal farmer with limited options. Now that Wealthy's best tricks have been trumped by newer apples, its prospects are about as good as Yahoo's.

Westfield Seek-No-Further

Aliases Westfield Seek-No-Farther; Seek-No-Further **Origin** Westfield, Massachusetts, 1700s. **Appearance** You will know it by its lenticels. Like Esopus Spitzenberg, for which it would make a fine body double, it is covered in a dusty red firmament through which its fat, irregular dots seem to swirl like the stars in Van Gogh's *Starry Night*. Like many rich-flavored apples, the flesh is yellowish. Sometimes there is a bluish bloom that can be shined away. **Flavor** Sweet and nutty. It displays high apple notes spiked with limey zest, vanilla, and enough astringency to either delight or deter you. This is nature's apple-walnut crisp. **Texture** Crisp and coarse with a dreamy snap. **Season** Mid-September to early October. Will keep for only a month or so. **Use** Best fresh. Turns mushy if baked. **Region** Rare, but treasured, among New England and Upper Midwest orchardists.

The Seek-No-Further encapsulates the saga of heirloom apples. A celebrated fruit in its hometown in the 1700s, it was propagated throughout the Connecticut River Valley towns of Massachusetts and Connecticut. The 1846 Committee of the New York State Agricultural Society wrote that "for many miles up and down, and round about that river, it is *the* apple, *par excellence*, of that locality; as much so as is the Newtown Pippin on Long Island, or the Esopus Spitzenburgh in Ulster. Whole orchards are planted of this fruit, and no where does it flourish in higher luxuriousness and perfection. It loves a warm, free soil; it is hardy, vigorous, and prolific. In flavor, it is excelled by few apples whatever for all household purposes."

Yet Westfield Seek-No-Further's excellence was closely tied to its home. Away from that warm, free soil, it did not grow red or tasty; in the South, it often russeted. By 1900 it had been abandoned by most commercial growers, and it fell into near-extinction. Now it is something of a cause célèbre in New England, where its virtuoso flavor and country-cute name give it Martha Stewart appeal. I think of Westfield Seek-No-Further as the ultimate cheddar cheese companion. A lump of cheddar, a tankard of cider, a few pickled whatevers, and a couple of Westfield Seek-No-Furthers, and you have the quintessential New England farm lunch.

BAKERS & SAUCERS

For all its excellence as a fresh fruit, not to mention its cultural and symbolic value, the apple soars higher still as our premier culinary fruit. For that, we can thank its unusual cellular structure.

While most fruits are packed tightly with thin-walled cells oozing juice, the apple is airy. Up to 25 percent of its volume is air trapped between its cells, and its cell walls are strengthened by plentiful cellulose and pectin. Cellulose is like the tiny two-by-fours that form the structure of the cell wall, and pectin is the glue that holds them together. That keeps apples dry and long lived, until those cells are broken by jaws, cider press, or heat. Heat melts the pectin in the cell walls so the cellulose "timbers" fall apart, allowing the air to escape and the juice to evaporate out of the apple.

Some apples have more air or cellulose than others. The tighter an apple's cells are packed together, the less air there is to escape during cooking, and the less the apple will cook down. You can tell how dense an apple is by closely examining the cells in its flesh, but you can also tell simply by hefting it; the heavier it is for its size, the less it will reduce when cooked.

An apple with a lot of cellulose in its cell walls will be firm and crunchy—sometimes too crunchy when fresh—and that is the key to a good baking apple, because some of its cellulose will remain intact when cooked. (A good sauce apple, like McIntosh or York Imperial, has less cellulose, which allows its cells to dissolve into a smooth puree.) Acid also helps by preventing pectin from melting when heated, keeping the cell walls more resistant. This means that not only will an acidic apple taste more refreshing in a sugary dessert, it will also retain more firmness.

Modern dessert apples often make passable culinary apples, because they have strong cells for maximum crispness, but they have neither the density nor the acidity of a true baking specialist like Calville Blanc. If you have never tried a top-notch baking apple in your pies and tarts, you are in for a treat.

Belle de Boskoop

Alias Boskoop **Origin** Boskoop, Holland, 1856. **Appearance** Large and round, elongated and sometimes lopsided, you can easily see the belle in this russeted yellow-green apple splashed with orange-red paint. **Flavor** One of the tartest of apples, with intense citrus aromatics. As she mellows with age, her sweetness begins to shine. **Texture** Firm and a bit dry. **Season** September for tangy purposes, October otherwise. **Use** A strudel star. **Region** Rare, but she turns up in many cool-climate heirloom orchards on both sides of the United States.

The Netherlands' most famous apple, this beauty is tart and snappy, with an acid tongue and a rustic coarseness. Picture a ruddy barmaid in some nineteenth-century Holland tavern. New York's Beach called the Boskoop "rather too acid for an agreeable dessert apple," but I know those who strenuously disagree. In any case, there is clear consensus on Boskoop's culinary talents; she will win you over in pies, crisps, and especially strudel, where her firmness is divine and her zippy edge keeps things lively. As you'd expect from a Holland apple, Belle de Boskoop does better than most in damp locations.

Bramley's Seedling

Origin Southwell, Nottinghamshire, England, 1809. **Appearance** Hangs in the tree like a grapefruit, huge and flattish and pale green. Eventually streaks of red bloom on the sunny side. **Flavor** Unflinchingly tart and acerbic. One-dimensional, but it's a fine dimension. **Texture** Super crisp. Carrot crisp. Makes a sound when you snap a slice in half. Yet it goes limp upon cooking. **Season** October. **Use** Pies, pies, and more pies. The thick skin peels right off in a clean spiral. The huge size is very convenient—three apples make a pie. Gets greasy in storage, so make those pies soon. **Region** Rare in the United States; legion in the United Kingdom.

In 1809, a little girl named Mary Ann Brailsford took some seeds from an apple being sliced by her mother and planted them in some flowerpots in her house. To her delight, one of the seeds germinated and soon showed a surprising vigor. Outgrowing its pot, it was transplanted to the garden, where it continued to thrive. A few years later, it began producing apples unlike any ever seen in England before or since. They were huge, green, oblate, and very, very tart. They had a remarkably high percentage of juice and low percentage of dry matter, which caused them to melt and bind together—exactly what the English like in a pie. The house and garden were later purchased by a butcher named Matthew Bramley, who in 1856 allowed a local nurseryman to graft the tree and begin propagating it, in exchange for naming rights. By the turn of the century, Bramley's Seedling had become the British pie apple, and today Bramley's is the automatic choice of most British pie makers. (Sometimes they are even referred to as simply "cooking apples.") Bramley's Seedling apples are a £50 million industry, with their own trade association. Those grown in County Armagh, Northern Ireland, are particularly famed for their hyperacidic character, and in 2012 they achieved EU protected geographical indication status, like the best wines and cheeses.

Bramley's is the British answer to France's Calville Blanc, vying with it for world pie domination. Yet the two couldn't be more different. While the Gallic *pomme* holds up firmly through cooking—a quality valued by most French and American cooks—Bramley's wilts into a smooth, thick, and biting pulp, for a characteristically British pie. In *The Epicure's Companion*, Edward Bunyard wrote that "the best English apples by long training know how to behave in a pie; they melt but do not squelch," yet I have Bramley's pegged as a squelcher, and a fine one at that.

In 2009 Bramley's Seedling celebrated its bicentennial, with BBC specials and many festivities. The original tree, now 205 years old, soldiers on in its garden, producing bumper crops of fat green apples, one of the stateliest apple trees you will ever see, and probably the most famous. It is feted every October by the town of Southwell at the Bramley Apple Festival.

Calville Blanc

Aliases White Winter Calville; Calville Blanc d'Hiver **Origin** Normandy, France, 1598. **Appearance** The most famously ribbed of all apples, its five prominent lobes make it look more quince than apple. If you've wondered what those weird things are in Claude Monet's *Still Life with Apples and Grapes*, now you know. Shiny and deep green in September, it gradually turns yellow-pink in the sun. The snow-white flesh will not brown easily. **Flavor** Simple, acid, clean, lemony. **Texture** Hard and snappy. Very dense. Yet with an exquisitely fine grain. **Season** October and November for pies. December for fresh eating (if you must). **Use** If you find Granny Smith too sweet and flabby for your tastes, then Calville Blanc is the fresh apple for you. Others save it for perfection in galettes and tarts. **Region** Frenchmen wax poetic about Calville Blanc, yet it's hard to find even in Normandy. Its new home seems to be New England, where a few specialty orchards grow it in abundance.

Calville Blanc wrings the superlatives out of people, who rave about its elegant Victorian looks and its rich, vanilla flavor. I find it intriguing yet awkward looking, and one-dimensional in flavor. Still, that dimension—intensely lemony—combines with its unparalleled firmness to make Calville Blanc the queen of tarts. It has stood the test of centuries of French chefs, and no proper tarte Tatin should be fabricated without it. Any time you need a cooked apple to hang on to a bit of its raw soul, Calville Blanc should be the first thing you reach for. It also makes an incredibly tart cider, unbearable on its own but excellent for ice-cider blends, where it sets off the concentrated sweetness. Some of that tartness also survives in dried Calville Blancs. Raw, Calville Blanc is the kind of apple Thoreau famously described as "sour enough to set a squirrel's teeth on edge and make a jay scream."

Cortland

Origin Geneva, New York, 1898 (cross of Ben Davis and McIntosh). **Appearance** Large, red, Mac-like. Gets very red, with a bluish bloom. **Flavor** Takes the Mac flavor and pushes it even further. A delicious, Champagne sort of flavor. **Texture** Fine grained and juicy, but soft, soft, soft. Mushy. Will not keep. **Season** Best in September. **Use** Elegant flavor and mushy flesh point this apple firmly in the direction of sauce, where it is unsurpassed. Also good for salads because it is slow to brown. **Region** Centered in the Finger Lakes region, where it was born, Cortland has spread to the Great Lakes, northern New England, and Canada, due to its cold-hardiness.

The scion of two of the most famous (and famously flawed) apples to ever fall from a tree, Cortland manages to combine some of the best attributes of McIntosh and Ben Davis. Bred in 1898 by none other than S. A. Beach, author of the revered *Apples of New York* and main man at the New York State Agricultural Experiment Station, it was released to the industry in 1915 and was the first release from that granddad of ag stations, or any other. It has done rather well for itself, despite the fact that no one I know is crazy about it. The soft flesh makes it disappointing for fresh eating, unless you catch it early enough in September that there is still a hint of snap and tartness; after that, it is an excellent sauce apple and a decent choice for salads, where truly crisp apples can be too distracting. For years it clung to a spot in the top twelve for U.S. apple production but now the likes of Honeycrisp and Pink Lady have bumped it down to the number-fifteen spot, and a fall from the top twenty seems inevitable.

Glockenapfel

Alias Pomme Cloche **Origin** Switzerland, 1500s, possibly earlier. **Appearance** Unforgettable. Long, ribbed, boxy, truncate. Looks as if it got stretched. Canary yellow with one mango cheek. Snowy flesh. **Flavor** Nicely sweet and very tart, with an unusual flavor like allspice and raspberries that are just beginning to mold (which is not a bad thing at all). When unripe, it has a bizarre fiddlehead finish, but by December it is all lemon and berry lightness. **Texture** Crisp, juicy, good. Its coarse, al dente flesh is excellent to sink your teeth into. **Season** Can be used for baking in October, but if left late on the tree its sweetness will develop and its green flavors will fade. **Use** Strudel, and other tarts and pies where firmness is critical. **Region** German-speaking Europe. A real find in the United States.

Glockenapfel translates as "bell apple." This gorgeous and distinctive apple hails from strudel country, and is still popular in Germany, Austria, and especially Switzerland. Obviously, it can take serious winters. This is *the* strudel apple, and has been for hundreds of years. A very hard apple, it not only holds its delightful shape through all sorts of culinary manhandling, but will also keep well through the winter. It could be one *Martha Stewart Living* feature away from cult status.

Granny Smith

Origin Eastwood, New South Wales, Australia, 1868. **Appearance** Large, glossy, uniform green with soft, off-white speckles and a peened surface. **Flavor** A shriek of acid and chlorophyll. **Texture** Crunchy, hard, squeaky. Can be surprisingly juicy. Donate the thick skin to your compost bin. **Season** Does plastic fruit have a season? **Use** A good pie apple. **Region** Still big in Australia and the warmer sections of Europe and the United States. Available anywhere, anytime.

Everything about Granny Smith is GREEN. The skin color. The flesh color. The flavor. Even the juice has a green tint. The very soul of this apple seems unripe. Somehow that one-dimensional quality has propelled this apple to international stardom. In fact, it was the first apple to conquer the world, toppling market after market on the strength of its indestructibility. With that plasticine green sheen and open-ended shelf life, it is as close to artificial as a real fruit could be. Yet that unripeness is probably not what attracted sixty-eight-year-old Maria Ann Smith to the seedling tree she spotted in 1868 growing on the bank of a creek where she had dumped a rotten batch of French crab apples from Tasmania a few years earlier. That's because, when grown in a warm, sunny climate and left on the tree well into winter, Granny Smith turns mellow yellow, with both the acidity and the vegetal greenness much diminished. Yet few people ever experience that fruit. And perhaps few want to—there are many who ask for nothing more from an apple than sour crunch. For them, the Granny never disappoints.

In Australia, Granny Smith was popularized as a cooking apple in the early 1900s. It began to be exported in the 1950s, in time for the surrealist René Magritte to paint one in 1966 with the words *"au revoir"* scribbled across it, a reference to the Garden of Eden, although by choosing the advance guard of the international fruit juggernaut for his painting, Magritte may as well have been saying good-bye to the domestic apple industry. The painting wound up in Paul McCartney's house in 1967, just in time to become the inspiration for the Beatles' new company Apple Corps (pun intended) and its subsidiary Apple Records. Considering their worldwide trajectory, it seems fitting that the Beatles chose Granny Smith to spin in the center of their records and not Cox's Orange Pippin.

Granny Smith hit America about a decade after the Beatles, supplanting older varieties as the go-to culinary apple, and carving out a year-round spot in the produce aisle. Although now surpassed by Gala and Fuji in sales, it continues to hold its own as the world's top choice for pies. It won't dazzle you with interesting flavor, but if you leave that to the cinnamon and nutmeg, then Granny will give you a base of firm tartness to build on.

Grimes Golden

Alias Grimes **Origin** Brooks County, West Virginia, 1790. **Appearance** Truly, madly, deeply yellow. Most apples said to be yellow are actually yellowish green, but Grimes Golden is canary. Small, round, magical. **Flavor** A big blast of banana and licorice, some say coriander, balanced by a fine skein of acidity. Not terribly complicated, but awfully yummy. **Texture** Crisp, fine-grained white flesh that cracks on contact. The only strike against Grimes is its tough skin. **Season** September in the South and Mid-Atlantic; October to November up north. **Use** Fresh eating and cider. Extraordinary applesauce. Does not keep long. **Region** Still common in the South; occasionally found in New England and the Midwest.

The moonshiner's apple, Grimes was spotted by Thomas Grimes when he bought a West Virginia farm in 1802. There was once speculation that Johnny Appleseed might have planted the original seed, since he had established a nursery in the area, but he didn't pass through until 1796, by which time the tree would have already been about six years old. By 1802 the twelve-year-old tree was filled with golden orbs that were hard to miss. Soon Grimes was selling the apples to traders who carried them by boat down the Ohio and Mississippi Rivers to New Orleans, just as they did with Ben Davis apples. (One hopes the Grimes Goldens commanded a premium.) Grimes grafted a full orchard of the delicious apples, but Grimes Golden remained a local phenomenon until the second half of the nineteenth century, when its fame spread throughout the Ohio River Valley and Southern states. The original tree continued to bear superb crops for a century, until it fell in a 1905 storm, taking a full load of fruit with it. Today, a monument in Grimes Golden Park, Brooks County, West Virginia, on Highway 27, marks the location of that first tree.

Grimes's high sugar levels and prodigious production made it the darling of Blue Ridge Mountain cider and brandy producers. Though it has fallen out of commercial production, it continues to be a sentimental favorite. A true child of Appalachia, Grimes in the North doesn't always develop the yellow color or rich flavor it does in the South, though it keeps longer.

Although Grimes Golden is famed in its own right for its wonderful flavor and versatility, it is overshadowed by its seedling child, also of West Virginia, born along a fence row near a Grimes Golden tree in Clay County in the early 1900s: Golden Delicious.

Jonagold

Origin Geneva, New York, 1968. (Cross of Jonathan and Golden Delicious.) Appearance A big, round, unabashedly red apple with yellow freckles showing through. Shiny and handsome. Frankly, the one pictured here is untypical, but so snazzy that we couldn't resist. Flavor Strong sweet and sour juices dance together over your palate in perfect step. Not a particularly complex flavor, but a very satisfying one. Texture Much like a Golden Delicious. Crisp and juicy when first picked, then slowly tenderizing over several weeks. Tough skin. Season September to October. Does not keep well. Use One of the great baking apples, it stays reliably firm. Good out of hand, and possibly the best of all dried apples—tart, raisiny, and spicy, with a hint of ginger. Region Nationwide. Quite popular in continental Europe, though not in England.

Michael Pollan calls this graduate of Cornell University "one of the great achievements of modern apple breeding," and I can think of no other product of the breeding programs that so successfully combines the good traits of both parents. Jonagold takes Jonathan's cheerful bright redness and lays it over Golden Delicious's womanly roundness. It takes Golden's honey-scented sweetness and spices it up with Jonathan's flowery tartness. It's an even better baker than Golden, and a more reliable dessert fruit than Jonathan. In the Cornell University canon, this rates well above Cortland and Empire and neck and neck with Macoun. Although it has never achieved the popularity it deserves in its native land, Europeans took to it long ago.

Jonathan

Origin Farm of Philip Rick, Woodstock, New York, 1790s. **Appearance** Looks not unlike its parent, Esopus Spitzenberg, with a similar medium size and roundish shape, though Jonathan tends to be more uniformly red, without Spitz's mottled look. **Flavor** The best description for Jonathan is "intensely appley." It doesn't hit you with any exotic or surprising aromas, but it does seem to concentrate traditional apple goodness, almost like an apple pie. Early in September, it can be nicely tart and spicy. **Texture** Delightfully snappy in September, taking on a Mac-like meltingness in October. **Season** Firm and tart when picked in September, it should be used by October. **Use** Excellent eaten straight off the tree. A famous pie apple, but only when picked crisp. **Region** Still a major player in the Midwest.

Jonathan is the essence of the American apple. That spicy, uncomplicated, sweet-tart flavor that most Americans think of when they picture biting into an apple is what Jonathan is all about. This has not made it universally popular. The English apple aficionado Edward Bunyard held up Jonathan as the epitome of the base American apple: "The great bulk of Jonathans, Winesaps, and their kin are apples, it is admitted, and as roughage for the ever-hungry young they play a useful part, but it has not yet been my fate to find among them a fruit of quality or distinction." Then again, Jonathans are said to achieve full Jonathan-ness only in the American Midwest, so probably Bunyard never tasted a prime one. Many a Midwesterner swears by Jonathan, which is still a staple in the orchards of Michigan, Illinois, and Ohio, enough that it is jockeying with York Imperial for twelfth place in U.S. apple production.

The original tree sprang up on the Rick farm, outside of Woodstock, sometime in the 1790s. In 1826, a local lawyer named Jonathan Hasbrouck brought his friend Judge Jesse Buel, president of the Albany Horticultural Society, to see the apple. Buel fell in love with it and quickly spread its fame (along with its scion wood). The apple tends to stay small and misshapen in Northeastern soils, and thus could never supplant Baldwin, yet in the rich river valleys of the Midwest it grows large and sweet. The comfortingly fruity Jonathan flavor has led Jonathan to figure in many breeding programs. Its progeny include Idared, Melrose, King David, Jonagold, Jonafree, Jonamac, Mutsu, and Akane.

Kavanagh

Alias Cathead **Origin** Damariscotta Lake, Maine, 1790. **Appearance** Looks like a pointy cat's head or, sometimes, a large, brown sno-cone. The yellow flesh turns deep brown very quickly. **Flavor** Unique and complex. Extremely floral and minty, like ice wine or muscat. The initial impression is of strong sweetness, but then, just when things begin to feel cloying, Kavanagh veers toward a gentle gooseberry tartness. **Texture** Neither crisp nor mealy, it is pleasantly granular, almost like a granita. Medium dry, it generates just enough juice to keep things flowing. **Season** September to October. **Use** Eat fresh, make sauce, or, as John Bunker recommends, fry in bacon fat. **Region** Maine mid-coast. That's it.

In 1780 an Irishman named James Kavanagh arrived in Boston, seeking his fortune. He found it in Newcastle, Maine, where he built a waterpowered mill on Damariscotta Lake to serve the shipbuilding industry and settled in, building his own mansion in 1803. Somewhere along the way he planted a very unusual apple tree, likely derived from seeds he carried with him from Ireland. The tree bore large, delicious, russeted apples shaped like a cat's head, minus the pointy ears, and that is what the apple was soon nicknamed. Kavanagh grafted many more of his trees on his land and shared scion wood with his neighbors. Soon the whole Newcastle area was filled with Kavanagh apple trees. Yet by the twentieth century, the Kavanagh was thoroughly forgotten.

Enter a ninety-year-old University of Maine professor named George Dow, who retired to the Damariscotta Lake area in the late twentieth century and became obsessed with the Kavanagh apple after discovering what he thought was the last surviving tree. Around 2000, Dow showed the tree to John Bunker, the Maine apple detective and owner of Fedco Trees. Bunker was immediately taken with the apple and its unusual form, which he knew was a classic shape for Irish apples. "It made a great applesauce—it fluffs up beautifully. But it was best known as a frying apple. They used to fry it in bacon fat."

After much detective work, Bunker found three additional Kavanagh trees along Maine's mid-coast, where it had once spread on the power of its gastronomic delights. In Freeport, he found a big, healthy, prolific Kavanagh in a parking lot. "On one side was this kayak sales center. On the other was this auto repair shop." The tree was in a patch of grass, possibly saved by growing right on the property line. "It was dropping fruit on the hoods of cars in both parking lots. I went to the owner of the auto repair shop and said, 'Please don't ever cut down this tree.' He said, 'I would never cut down that tree. I love that tree.'" Bunker now has a collection of fruiting Kavanaghs on his property, and has sold hundreds of trees through Fedco. Long live the Cathead.

Maiden's Blush

Origin Burlington, New Jersey, 1817. **Appearance** A small, winsome apple, with waxy yellow skin freckled with green and a bright fuchsia blush on its sunny cheek. **Flavor** Full of bite off the tree, but mellowing to a sweet and perfumed spiciness full of honey and star anise. **Texture** Crisp enough when picked, it feels light for its size, dry, and insubstantial. The paper-white flesh is very smooth and fine grained. It has unobjectionable, thin, waxy skin, like a succulent. **Season** September. Pick early if you want tartness. **Use** The quintessential dried apple. **Region** Once an important apple in the Mid-Atlantic states, now rarely seen.

This graceful apple originated on the New Jersey farm of Samuel Allison and became a staple of the home orchard in most Mid-Atlantic and southern Appalachian farmsteads. By 1838, it was already "a very popular apple in the Philadelphia market," according to William Coxe, America's first apple expert, who described its flesh as "remarkably light and fitted for drying, for which it is preferred to any apple of the season." Being low in juice, it dried easily, and the flesh stayed an attractive white. Dried apples were an important way to extend the harvest in nineteenth-century America, and they still make unique desserts. The Maiden's Blush stayed popular for its drying ability for a century, and now survives purely on its looks.

McIntosh

Origin Dundela, Ontario, Canada, 1811. (Snow/Fameuse seedling.) **Appearance** The classic Mac is a shiny yin-yang swirl of green and cherry red, very charming en masse in a bin or basket. Some get a little stripey. The lower half in particular gets quite speckled with green dots. Medium-size, round, and somewhat flattened, Macs always feel a little soft and waxy. The flesh is white. **Flavor** Macs taste cidery. It's an edge-of-ferment flavor, like soft strawberries, what the old books used to refer to as *vinous*. They are savory and tart, with the slightest edge of curry lingering on your lips, sometimes quite noticeable a few moments after you've eaten the apple. **Texture** Picked early in the season, Macs can be plenty crunchy, but by October the soft flesh quickly melts away, leaving you gnawing on the interminable skin. **Season** September. **Use** Superb for fresh cider and for sauce. Love it or leave it raw. **Region** As New England as the Red Sox. In Canada, Macs account for half the apple crop.

Those who know the Mac as an iconic part of fall in New England are surprised to learn that it's a Canadian apple, and a relative newcomer. The first Mac was found as a wild seedling by John McIntosh as he cleared land on his Canadian farm near the U.S.-Canada border. He transplanted it near his farmhouse and the apple tree became locally famous. McIntosh sold seeds and seedlings to his neighbors and others passing through, only to discover a few years later that the young trees did not make Mac apples. Not until 1835 did someone teach him the art of grafting. The original tree died in 1910 at the age of one hundred plus. At that time, the McIntosh was still merely a regional understudy, scrabbling at the edges of the industry like a little mammal waiting for the dinosaurs to croak. It didn't even merit a listing in volume 1 of *The Apples of New York*, published in 1905, though volume 2 noted that the obscure apple "is regarded by many as one of the most promising varieties of its class for general cultivation," praising the "very tender, perfumed and delicious" flesh but cautioning that "because of its lack of firmness it is less suitable for general handling." This has always been the knock against the Mac: great taste, deflating mushiness. Actually, there are many knocks against the Mac: mushy flesh, skin like a rhinoceros, small size unless aggressively thinned, and terrible susceptibility to scab, the apple grower's worst nightmare.

The Mac would never have gone on its sixty-year run of dominance were it not for two developments in the twentieth century. One was the invention of commercial sprays, which eliminated the Mac's scab problem. The other was the great freeze of 1934, which killed the vast Baldwin orchards of New York and New England. Growers looked for a new red apple that was extremely cold-hardy, so this would never happen again, and the Mac filled the bill. It loved the cold, looked good, set an abundant crop of fruit that ripened simultaneously every year, and possessed that curious flavor that managed to worm into the brains of several generations of New Englanders and curl tight around the memory center.

Still, we eaters of apples care less that the Mac sets a perfect crop every year than that it sets a perfectly thick peel every year. My childhood was littered with a trail of inedible Mac skins. They wouldn't be quite so annoying were it not for the melting flesh. The contrast is like eating a lovely, creamy custard wrapped in nori. And now that sprays are not so cool, the Mac may one day be no more famous than Fameuse.

Newtown Pippin

Aliases Newton Pippin, Yellow Newton Pippin, Green Newton Pippin, Albemarle Pippin **Origin** Newtown (now Elmhurst), Queens, New York, 1720. **Appearance** An unassuming smooth green apple the color of a Florida field tomato, with a few white dots. Medium sized and utterly forgettable. Over the winter, the green skin develops a jaundiced edge or sometimes a wan pink blush. Often, Newtown Pippin is mottled and lopsided. Not a looker. **Flavor** Somewhat sugary and very acid, with a bracing, lemony flavor and a green-tea note from the skin. Clean and spare, like a Granny Smith. **Texture** Very firm and crunchy. **Season** Pick in October or early November, but wait to eat. Erik Baard, who founded newtownpippin.org in New York City, calls the freshly picked Pippin "a starchy, puckering ball of astringent un-fun." Those qualities are actually what make the Pippin one of the best keepers of any variety. In fact, like fine wine, it needs to breathe for a while before its aromas open up. Over the winter, the tannins and acids mellow, the starches turn to sugar, and magical new aromas awaken. **Use** Good out of hand, and superb in pie, where its shape holds up beautifully. A premier cider fruit. **Region** Grown commercially in Virginia and California. Making a comeback in New York City. A presence in many heirloom orchards.

Like Forrest Gump, the Newtown Pippin has managed to intersect with an improbable number of historic personages and places over the course of its career, and has shown a knack for effortless success at whatever it was called upon to do. Let us start at the beginning: Newtown, New York (now Elmhurst, in Queens), the orchard of Gershom Moore (vicinity of Broadway and Forty-Fifth Street), 1720: A seedling is born. It soon produces stunning large, green apples with superior flavor and unparalleled keeping abilities. This Newtown Pippin is named and widely grafted—making it the only apple variety to have originated within the five boroughs of New York City—and it becomes a staple of the country's first commercial nurseries on Long Island. When Benjamin Franklin was staying in London in 1759, he received a gift shipment of Newtown Pippins from Long Island. The British fell in love with the apple and quickly imported scion wood from Long Island, but the apples always stayed small and crabby when grown in England (except when they were grown in glass greenhouses, which proved successful but impractical). This began Pippin Mania I in England, where imported Newtown Pippins sold for insane prices.

By the 1750s, Newtown Pippins were being grown not just on Long Island but throughout New York and Pennsylvania. That was where Thomas Walker discovered them. Walker, who owned a fifteen-thousand-acre plantation in Charlottesville called Castle Hill, was serving as commissary for General Braddock's British army in the French and Indian War. In 1755, Braddock's army was routed (and Braddock was killed) while attacking a French fort in the location of modern-day Pittsburgh. While the remnants of the army (led by a young George Washington) sought winter refuge in Philadelphia, Thomas Walker loaded up his saddlebags with cuttings from Newtown Pippin trees and headed back to Virginia. Within a few years, everyone in Charlottesville's Albemarle County was growing Newtown Pippins, and people soon recognized that in the rich, loamy soils of the Virginia Piedmont, the Pippins grew larger and sweeter than they ever had on Long Island. Walker served as guardian to Thomas Jefferson after the eleven-year-old Jefferson's father died in 1757, and may well have introduced Jefferson to his first Pippin. Jefferson planted Pippins at Monticello in 1778 and had more success with them than any of his other apples, eventually planting 170 trees. While serving as ambassador to France, he exulted in a letter to James Madison that "they have no apples here to compare with our Newtown Pippin." George Washington also adored the apple and grew it at Mount Vernon.

Through the early 1800s, the Pippin continued to garner adulation

on both sides of the Atlantic. The horticulturalist William Coxe was deeply in love with it: "The finest apple of our country, and probably of the world." The British, in a desperate attempt to protect their tiny domestic apple industry, had slapped a tariff on imported apples, but in 1838 America's ambassador to the United Kingdom, Andrew Stevenson, had two barrels of Newtown Pippins shipped from his Virginia home. He slipped a basket of them to eighteen-year-old Queen Victoria, then in the first year of her sixty-three-year reign. "They were eaten and praised by royal lips and swallowed by many aristocratic throats," Stevenson's wife recounted. The young monarch was so besotted with the green apples that she waived the tariff on them alone. Thus began Pippin Mania II. Pippins sold for three times the price of other apples, and experts warned of inferior apples being passed off as Newtown Pippins in London markets.

At that time, New York's Hudson Valley and Virginia's Albemarle County were competing to export their Pippins to England. The largest apple orchard in the world was that of the "Apple King" Robert Pell, whose twenty-five thousand trees in Esopus, New York, were predominantly Newtown Pippins. The export trade was handled out of New York, which was sourcing its apples more and more from Virginia, and taking the largest cut of the profits. But, perhaps with Stevenson whispering in their ears, the British tastemakers insisted on the Albemarle Pippins, and Virginia growers learned to cut out the New York middlemen, shipping their "Albemarle Pippins" directly out of Baltimore. That name has become standard in Virginia.

Pippin Mania petered out after World War I, though in 1929 Edward Bunyard wrote, "Of the imported apples some come into the highest rank, Newtown Pippin is supreme in its own class, the aroma of pine and the refined flavour are worthy of all praise." But the variety nearly disappeared from the East Coast, which was switching over to more productive apples. It fared better on the West Coast, where it had been introduced in the 1850s. Although it eventually lost its spot as the hard, green apple on store shelves to the more prolific Granny Smith, Newtown Pippin still provides the pizzazz in Martinelli's Sparkling Cider.

Now, a movement is afoot to return the Newtown Pippin to semi-official status as New York City's apple. It has been celebrated by chefs such as Peter Hoffman and Dan Barber, recognized by Mayor Bloomberg, and championed by Slow Food NYC, which acquired Newtown Pippin scions from Virginia and provided them to local orchards. The Pippin is probably the only apple with a nonprofit devoted to it—newtownpippin.org. Founder Erik Baard has planted Pippins at schools, churches, parks, and other public spaces throughout New York City. When they begin to fruit in the next few years, the green apple could become an iconic sight in the Big Apple.

The apple is also once again covering hillsides in the Virginia Piedmont, this time because of its rediscovery as a world-class cider apple. Albemarle Pippin makes a stunningly clear, light, and tart sparkling cider, very much like a blanc-des-blancs Champagne, and it can now be found at half a dozen cideries in the Charlottesville area—including Castle Hill, the site of Thomas Walker's first grafted Pippin.

Northern Spy

Origin East Bloomfield, New York, 1800. **Appearance** There is always something pink about the shoulders of Northern Spy. At a passing glance, you might call it red, but in a lineup of truly red apples its pinkness becomes apparent. It can get quite large and handsome, with ribbed sides and ripped shoulders. A deep stem depression makes it look even more as if its shoulders are rising up. Handsome pinkish red stripes drop over those greenish shoulders, cascading to a pale yellow eye. Like Red Delicious, Northern Spy sits on five knobs. The flesh ripens to a butter-yellow. **Flavor** One of the great combinations of syrupy sweet and screamingly tart. That intensity, combined with the size, can make Northern Spy an all-consuming experience. When you bite into it, there is a fleeting salvo of something rich and fruity, then the blast of sharpness arrives with a classic apple nose. At the end, your lips feel etched in acid. **Texture** Crisp and tender. Perfect. An apple's apple. You are too busy enjoying the rivulets of sweet-tart juice to notice how delightfully thin the skin is. Northern Spy is fun to eat. **Season** Pick in September for pie, October (after the frost) for fresh eating. It bruises easily, but still keeps with those bruises. **Use** An apple that can do it all. Great cider, great pie. It won't cook down to nothing. A big, fresh Northern Spy will keep a sour-loving kid occupied for an hour. **Region** Still common in the Northeast and Great Lakes region. Look for the plaque in East Bloomfield where the original tree stood.

If I had to live on a desert island with only one apple variety, it might well be Northern Spy, which I think of as the Calville Blanc of the Northeast: A big, firm, sharply acidic, superb winter pie apple. It ferments into a better hard cider than Calville Blanc—clean and crisp, with a steely edge of acidity running through it.

The name has caused a century of head-scratching. Some have proposed that it was originally Northern Pie. (After all, it does make fine pies.) Others, Northern Spice. (It is rather spicy, now that you mention it.) But our best evidence comes from an 1853 letter from Rochester, New York, to a gardening magazine, to wit: "In reply to Mrs. B. who inquired about the naming of the Northern Spy apple, everybody here knows it was named for the hero of that notorious dime novel *The Northern Spy*, but no one will come out and admit it." Understandably so. The anonymous 1830 novel *The Northern Spy* is about an abolitionist straight out of Quentin Tarantino, who established an underground railroad from Virginia to Canada. Posing as a slave catcher, he killed true slave catchers in ambushes, before himself dying in a battle. It was claimed by some that *The Northern Spy* was a work of propaganda that helped incite the Civil War. Considering the sensitivity of the topic of slavery in 1853, no wonder no one would fess up to the association, but once you know the connection, there's no denying that the apple fits. With its hulking shoulders and its acerbic temperament, Northern Spy is as formidable an apple as you'll find, and there is a cold-climate abolitionist's severity in its mien.

Although the Spy was born in 1800, it took its own sweet time coming into its prime. It can take ten years or more to fruit, and the original tree had already been nibbled to death by mice before ever bearing the first Northern Spy apple. Fortunately a brother-in-law of the original owner had already planted some sprouts of that tree (on a hunch, perhaps?), and around 1830, just as *The Northern Spy* was being secretly passed from one abolitionist's hand to the next, the apple was building a following throughout the Finger Lakes. It proved notoriously finicky about its terroir. Plant it in a wintry region on rich, well-drained slopes and you'll get a great apple; elsewhere, not. By 1900 it was the number-three apple in the Northeast, behind Baldwin and Rhode Island Greening but often superior in flavor to both. Only its ten-year Hamlet phase and its tendency to bruise in transport kept it from being the great apple of the Northeast—although many an Upstate octogenarian will insist that it is.

Northwest Greening

Origin Waupaca County, Wisconsin, 1872. (Cross of Golden Russet and Alexander.) **Appearance** Huge, but light for its size. Smooth, shiny, relentless green, like a gigantic Granny Smith. It has russet at the stem, sometimes a lot, and a few brown lenticels. The eye is wide open; you can look deep into the apple's core. **Flavor** Mildly sweet. One-dimensional. **Texture** Very coarse cells and dry, cardboardy flesh make this apple unappealing fresh. **Season** October. **Use** Bake or dry. **Region** Colder sections of the Midwest, especially Wisconsin.

It is difficult to distinguish Northwest Greening from Rhode Island Greening in appearance, but in appearance only. *The Apples of New York* pulled no punches in its 1905 description: "The fruit has a serious fault in that the flesh within the core lines is apt to be corky and discolored. It cooks evenly and quickly and when cooked has a fine yellow color but is not of high flavor or quality being much inferior in this respect to Rhode Island Greening." It's not even a good keeper. But it is extremely hardy, and thus was adopted in northern parts of Wisconsin and other places that couldn't grow Rhode Island Greening. I suspect that this huge apple was often dried, since the cottony flesh is already halfway there. Worth a taste if only to remind yourself that preindustrial homesteaders endured crappy apples, too.

Porter

Origin Shelburne, Massachusetts, circa 1800. **Appearance** At first glance an unmemorable green apple of medium size, the Porter, upon closer inspection, fascinates. On the side exposed to the sun, you'll find a soft terra-cotta wash, spattered with dull white spots, suggesting age spots on tanned skin—beautiful, but the sort of beauty that is lost on the young. Later, it turns bright yellow, with blotches of red on the sunny side that make it charming to all ages and dispositions. **Flavor** Middle-of-the-road appleness, fairly tart but not complex. The old books all rave about Porter's "agreeably aromatic" flavor and "fine dessert quality," but I find it to be a bit of a journeyman apple. **Texture** Fine grained, crisp-tender, and juicy. The excellent skin is tender and tasty. **Season** August in the South, early September in New England. **Use** Passable for eating fresh, it holds up great in pies, crisps, and when fried. The most famous of all canning apples. **Region** Eastern New England orchards and occasional heirloom specialists farther down the East Coast.

From a tree raised by the Reverend Samuel Porter outside Boston around 1800, Porter became "a great favourite in the Boston market" by the mid-1800s. In its favor were an early ripening date—before the masses of Baldwins and Roxbury Russets and Rhode Island Greenings hit the market—a pretty color, and a reputation for culinary excellence. In fact, it was the only apple mentioned by name in Bostonian Fannie Farmer's original 1896 cookbook, which featured a recipe for Canned Porter Apples. Clearly the apple made it to the New York market, too, because Henry Ward Beecher, the pastor of Brooklyn's Plymouth Church (and a product of Connecticut) asked in an 1862 essay, "Who would make jelly of any other apple, that had the Porter?" Yet Porter's pale, delightfully thin skin, which is partly what makes it attractive as an eating apple, bruises readily. The apple never shipped well. On top of that, it ripens unpredictably, meaning it needs to be checked every few days by someone, and it tends to produce fruit of different sizes. And, of course, it isn't red. All those factors kept Porter from ever becoming a successful commercial apple, though it was spread throughout the country by home orchardists, who discovered that the Porter did as well in the South and Midwest as it did in Boston. In the twentieth century, as the orchards that had supplied Boston with local fruit disappeared, the Porter went with them.

Pound Sweet

Alias Pumpkin Sweet **Origin** Manchester, Connecticut, early 1800s. **Appearance** This huge, green globe sets itself off from other varieties by size and by staying pure green, Granny Smith green, even when ripe. It is smooth skinned and ribbed, with five obvious sides. **Flavor** There is not a hint of acid in Pound Sweet, which give it, to most people, a very un-appley flavor. Instead, what comes across is a lightly caramelized sort of root-vegetable sweetness. **Texture** Coarse and soft. **Season** September. Use quickly. **Use** Neither a fresh eater nor a pie apple, but its size and sweetness lend it to baking whole or using in sauce. Famous for apple butter. **Region** Still found on New England and Midwestern homesteads. Rare in commercial orchards.

If you have a hankering for a batch of quince stew, this is the apple for you. As *The Apples of New York* put it in 1905, "By many it is esteemed as one of the best sweet apples of its season for baking and for canning or stewing with quinces, but generally it is not valued for dessert because it is rather coarse and has a peculiar flavor." Well, yes, it is rather coarse, but there's nothing peculiar in the flavor, unless you consider cream soda peculiar; it's sweet, delicious, and vanilla scented. That, along with its size, made it the go-to apple for a small Midwestern apple butter industry in the nineteenth century. Brad Koehler, who makes some of the country's finest ice cider at Windfall Orchards in Cornwall, Vermont, considers Pound Sweet's gilded flavors an important part of his blend. Small children are drawn to Pound Sweet, which practically glows like a luminescent green toy, and they stick around for the easygoing sugars. Adults have trouble embracing the soft, compliant apple. This was the exact kind of apple the English food writer Morton Shand—who favored the crunchy, tart apples of his homeland—had in mind when he sneered at "all those flaccid-fleshed pumpkins...that the French rave over."

Rhode Island Greening

Alias Greening, Rhode Island **Origin** Newport, Rhode Island, 1640s or earlier. **Appearance** The definitive big green apple, with a splash of sienna on the sunny cheek. Smooth skin, a little waxy, with a smattering of tiny brown triangular dots. **Flavor** Sweet and very tart, with a nice grape juice flavor. Avoid the skin, which has a nasty vegetal taste. **Texture** Crisp and crunchy. A serious apple. Coarse flesh, but not as coarse as most large apples. Juicy enough to keep things moving. When baked, it becomes tender but not mushy. **Season** September to early October. Will keep for several months if you refrigerate right away. **Use** *The* New England pie apple. Good fresh if you like very tart apples. One of the best dried apples, it develops a nice toothy texture and maintains its zing. **Region** Northeast; now uncommon.

Rhode Island Greening was the Granny Smith of the 1600s, 1700s, 1800s, and early 1900s, not being vanquished by the Blunder from Down Under until the 1970s. Rhode Island Greening was the premier culinary apple of the Northeast, second in production only to Baldwin. The two worked as a team, kind of like Red Delicious and Granny Smith: one sweet red apple for fresh eating, and one tart green apple for cooking. Many orchards split their land between these two varieties. Because the Greenings were picked slightly earlier than the Baldwins, they also made for an efficient combo in the orchard; the same crew could move on to the Baldwins once they were done with the Greenings. Rhode Island Greening's flavor and baking qualities are so superior to Granny Smith that many apple people hope it will become the Granny Smith of the twenty-first century, too.

The original Rhode Island Greening tree grew in Green's End, a village near Newport, where a fellow by the name of—yes, you guessed it—Green kept a tavern. Guests at Green's Inn became enamored of the huge, green, tart apples he served from the tree in his yard, and many asked for cuttings. Soon the fame of the "Green's Inn apple" had spread throughout New England. (Making it probably the likeliest candidate for the Tree that Ate Roger Williams; see Roxbury Russet for the full story.) So many cuttings were taken from the original tree that it eventually died the death of a thousand cuts, but by then its clones were taking the colonies by storm. It has always seemed suspiciously tidy to me that our most famous green apple began with a Mr. Green, but all the historical sources seem to agree on this.

Some of my greatest pies have been made with Rhode Island Greenings. The slices are huge and they hold their shape beautifully, fostering the high, lumpen, bountiful, so-many-apples-that-it's-busting-out-at-the-seams look. The tartness carries the pie. You wouldn't choose it purely on flavor, but as our ancestors knew, when it comes to pie, size, texture, and ease of processing can be more important criteria.

Get Rhode Island Greening early in the fall, when it is still assertively tart. It should be green as grass. Once the sienna hues creep into the skin later in the fall, it will make for sweeter fresh eating, but a wimpier pie. Get it in the north country, too; in the South, it never develops good flavor and tends to drop unripe from the trees. (Back in 1908, the USDA didn't sugarcoat it: "In most southern locations, it lacks nearly all points of merit.")

Rhode Island Greening trees are famously long lived, and they develop the wizened look of ancient fairy-tale trees as they age. In some of the more stately New England zip codes you can still spot grand specimens that are pushing 150 or even 200 years old.

THE
APPLES
OF
NEWYORK
BEACH
VOL. I
THE
APPLES
OF
NEWYORK
BEACH

Rome Beauty

Alias Rome, Gillett's Seedling **Origin** Rome Township, Ohio, 1817. **Appearance** Stunning apple with Hollywood appeal. Its smooth skin, high color, tight curves, and cute freckles mark it as the All-American Apple Next Door. **Flavor** Smells a bit like roses. Mildly sweet, not acidic, and incredibly perfumed. **Texture** Wow, that's some tough skin. The thick, mealy flesh is dry and unappealing. **Season** September to October. **Use** Cooking and drying. Keeps until spring. Awful fresh. **Region** Midwest, especially Ohio.

Like Cinderella, the Rome Beauty was never intended to be anything more than an invisible helper. When apple shoots are grafted onto rootstocks, the rootstock plays no role in the tree except to pump water and nutrients from the ground. It's been selected for its skill as a wet nurse, never a bearer of its own fruit. Whenever a rootstock sends up a shoot, it is quickly lopped off. Yet along the banks of the Ohio River in 1817, Joel Gillett yanked a shoot off the rootstock of one of his newly acquired apple trees and handed it to his son Alanson, who planted it without ceremony, probably thinking he might get some livestock feed out of it. Instead, the little tree began producing some of the handsomest apples the world has ever seen. The Rome Beauty was born (that's Rome, Ohio), though the flooding Ohio River took the original tree in 1860.

By the late 1800s, Rome Beauty was a commercial star, competing with Ben Davis in the Ohio River Valley, New Jersey, and some points south. Like Ben Davis, its thick skin and general starchiness made it a superb keeper, and it was once ubiquitous in general stores. As recently as 1995, Rome was the number three apple in America, ahead of all save the Delicious duo, with Mac, Fuji, and Granny Smith nipping at its heels. It tastes awful when fresh, but that dry gumminess is exactly what made it a famous cooker. In fact, this beauty was long known as "Queen of the Baking Apples" and "Baker's Buddy." Peeled and concentrated in a pie or crisp, it blossoms.

Smokehouse

Alias English Vandevere **Origin** Lancaster County, Pennsylvania, early 1800s. **Appearance** Medium size, squat and green, with a rusty wash on the sunny cheek, turning bright red late in fall. Large, raised lenticels scatter the surface. **Flavor** Very sweet and cidery, with malty caramel notes. Low acidity exaggerates the sweetness. **Texture** Firm and very dense, making it wonderfully substantial in pies and crisps, but a bit dry and mealy for topnotch fresh eating. **Season** September. The density is misleading; it does not keep. **Use** A good cooking apple, Smokehouse caramelizes nicely, though it benefits from a squeeze of lemon juice. When young, it makes fine fresh eating. **Region** Smokehouse loves the clay soils of the southeast Piedmont and Mid-Atlantic.

Do not expect a smoky flavor from this apple. The original tree grew up near the smokehouse of William Gibbons and was named accordingly. This variety has many strong and knowledgeable admirers, including the Maine apple collector John Bunker ("I like its subtle flavor right off the tree") and the seventh-generation Virginia orchardist Tom Burford (who lists it among his Top 20 Dessert Apples), but you have to catch it in that brief window before it loses its zest. Keep your eyes peeled at pick-your-own orchards, and maybe you, too, will have a Smokehouse moment.

Snow

Alias Fameuse **Origin** Quebec, 1600s. **Appearance** Like a Mac the size of a racquetball, splotched with fairly equal red and green sides. Full sun causes the Snow to burn brilliant crimson. The striking contrast of the lipstick-red skin and the whiter-than-white flesh feels like something out of a fairy tale. Sometimes the flesh is flecked with red—more Brothers Grimm than Disney. **Flavor** Mac-like but more tart and delicate, with a floral hint of white chocolate on the finish. **Texture** Tender when fully ripe, with unfortunately thick skin. **Season** Crisp and tart in early September; sweet and melting by late September. **Use** Good fresh, the sooner the better. Makes famously fluffy sauce, and excellent sweet cider. **Region** Thrives only in Canada and northern New England.

Although many authorities list this apple under Fameuse, I prefer its other name for the Snow White connotations. Like details in fairy tales, the Snow seems quietly and inscrutably significant. It has the unusual habit of coming fairly true from seed—Snow apples don't fall far from the tree. (Although the McIntosh, its most illustrious offspring, fell just far enough to eclipse the Snow in commercial appeal: Macs produce bigger fruit more consistently.) That quality, along with its delight in boreal climates, allowed it to conquer Quebec along with the French settlers of the 1600s. Snow apples appeared in the gardens of every *seigneury* from Quebec City to Montreal, and by the early 1700s the Snow had become *the* apple of French Canada, from the mouth of the St. Lawrence to the Great Lakes. Undoubtedly many of those trees were planted from seed, since few settlers knew the art of grafting at the time. (The experts argue over whether the first Snow originated in Canada or France, but since the variety is almost unknown in Europe, and poorly adapted to the climate there, my money is on Canada.)

In 1731, French troops followed the Richelieu River from Montreal down into Lake Champlain and established a fort at a narrows on the lake, which they named Crown Point. This was the first European settlement in Vermont. The idea was to keep the British out of Canada, which worked for a while. A thriving French settlement grew up around the fort, complete with extensive orchards of Fameuse apples. In 1759, with the French and Indian War at a fever pitch, the French were forced to abandon the fort, but the trees remained, and Vermont's settlers soon grafted scions from these trees throughout the Champlain Valley, which continues to be a Snow stronghold.

South of this region, the Snow turns too soft too fast, and never develops its fetching color or refreshing acidity. In the right place—the Champlain Valley, the St. Lawrence Valley, and other cold spots—it can be really good, and was an object of desire back when people valued tender apples more than crisp ones. "No orchard in the north can be counted as complete without this variety," said the 1865 Yearbook of the United States Department of Agriculture, which also praised the white, fluffy sauce it made.

Spokane Beauty

Origin Walla Walla, Washington, 1859. **Appearance** A huge (two pound!) flattish apple with a green background washed in burnt sienna. The inner flesh is white, sometimes tinged green. **Flavor** Winesap-like, with dilute sweetness cut by assertive acidity and a mown-grass herbaceousness, like Sauvignon Blanc. **Texture** Crisper than most large apples. Melting. Very nice to eat. **Season** October. **Use** Makes wonderfully zippy sweet cider. Fresh, it's a meal for two. The size is convenient for drying and cooking. **Region** Northwest.

This Brobdingnagian apple is probably the best to hail from Washington State. It won first prize at the Spokane Fruit Fair in 1895, and repeated as champion in 1896, yet its fame has not spread beyond the Northwest. That's perplexing, because there's nothing quite like the Spokane Beauty. It may well be the largest apple known, and unlike most giant apples, which tend to get soft and flabby, Spokane Beauty stays tight and tart. Put this high on your home-orchard list.

Stayman

Alias Stayman Winesap. **Origin** Leavenworth, Kansas, 1866. (Winesap seedling.) **Appearance** A handsome antique-looking apple, with a yellow-green background (turning totally yellow by October) nearly covered in a wash of dull carmine, like an old barn. The slightly conic fruit sometimes leans off-center, like a clay pot that lurched to one side before drying. The apple is slightly ribbed and lightly speckled with white dots. **Flavor** Like its parent, Stayman stays tart and vinous. In the north, it tends to stay astringent and unripe, about as much fun to chew on as a timothy stem, but in the South the acid is moderated by good sweetness. **Texture** Although not as crisp as a modern apple, Stayman has nicely snappy flesh, and the thin, al dente skin is a plus. **Season** September to October. **Use** Good fresh for tart apple lovers. Stays firm, tart, and shapely in pies and tarts. Makes delightfully spirited fresh cider. **Region** Still a common sight at roadside stands throughout the South.

In 1866, Dr. Joseph Stayman planted a number of seeds from his Winesap tree around his Leavenworth home. Nine years later, one of those trees produced such promising fruit that he sent samples to several experts. By 1890, Stayman's Winesap had been discovered, and by 1895 the powerhouse nursery Stark Bro's was popularizing it. Although Midwestern by birth, Stayman found itself in Appalachia and the Piedmont of Virginia and North Carolina, upon whose soils it developed its distinctive high flavor. It has been a big apple throughout the South ever since, considered superior in flavor to its famous mother. Southerners get misty-eyed over Stayman's memorably winey tartness, but commercial growers do not love the apple's tendency to split open if heavy rains come before harvest. Stayman is still among the top 20 commercial apples, but just barely. Production is just a sixth of what it was thirty years ago.

Twenty Ounce Pippin

Alias Oxheart **Origin** Unknown, early 1800s. **Appearance** One of the biggest and greenest of them all. Twenty ounces seems an underestimate. The huge ones can get quite conic, almost strawberry shaped. Yet the apple seems light for its size. **Flavor** Ho-hum. Mild and tart. The skin has a strong, green chlorophyll flavor, like green tomato. **Texture** Soft and dry. Mealy. The skin is fairly thin and pleasant. **Season** September to October. **Use** People kept these humongous, dry apples around for one purpose only, and it wasn't for eating fresh. Big and easy to use, they make lots of sauce or dried apples. The size makes them convenient for pies, but they don't stay firm when cooked. **Region** Northern states, from Midwest to Maine.

This apple's longevity rests almost entirely on a case of mistaken identity. There is a great old apple from Cayuga County, New York, known as Twenty Ounce. Many an apple enthusiast, commercial as well as amateur, thought he was raising that tree, eagerly looking forward to the tasty red fruit, only to find after a few years that he had been cuckolded by the Twenty Ounce Pippin and its large, tasteless apples. 1905's *Apples of New York* refers to the fruit as "attractive in appearance, but second or third rate in quality." In the 1800s, there was a failed movement to change the name of the good Twenty Ounce to Cayuga Red Streak, to clear up the confusion. But the confusion, like the Twenty Ounce Pippin itself, endures. Whatever its culinary shortcomings, it makes an impressive display.

Virginia Gold

Origin Virginia Polytechnic Institute, Blacksburg, Virginia, 1976. (Cross of Newtown Pippin and Golden Delicious.) **Appearance** The spitting image of its Golden Delicious mother. A golden globe of an apple with one salmon cheek, and sometimes a fine mesh of russeting. Solid in the hand. Very round. **Flavor** Sweet in October, with medium acid, like a much livelier Golden Delicious. The acid becomes quite prominent after a few bites. **Texture** Incredibly firm and crisp, with a solidity that works your jaw. The skin is rock hard and sharp. Peel! **Season** October. **Use** Excellent culinary apple in the fall and dessert apple in the winter. **Region** Southeast, especially Virginia mountains and Piedmont.

Virginia Gold is much sassier than her mom, full of tart rejoinders and capricious moods. You won't get bored with her so quickly. She gets those qualities from her dad, Newtown. Like Newtown, her flavor improves for several months in cold storage, and interesting pine notes emerge. Why this beautiful, delicious apple isn't better known escapes me.

Wolf River

Origin Fremont, Wisconsin, near the Wolf River, 1856. **Appearance** Massive, round, reddish pink, ribbed, like a Northern Spy on steroids. Inside, the flesh is alabaster, quickly browning when exposed. **Flavor** Hardly sweet, mildly acid, profoundly uninteresting, except right where the flesh meets the skin, where there is a nice hit of cherry. **Texture** Dry, a bit airy, almost spongy. **Season** Early fall for culinary uses. **Use** A baked apple specialist. Also good dried or in pies. **Region** Known nationwide for its size, but most popular in its native Midwest.

In 1856, a Quebecois woodsman named William Springer was traveling by wagon with his family from Quebec to Wisconsin, where he hoped to make a new life. Somewhere along the shores of Lake Erie, Springer stopped to buy a bushel of apples. The apples not only kept the Springer bellies full as they traveled, but they continued to feed the family after Springer planted some seeds from those apples on the site of his new farm, along Wisconsin's Wolf River. By 1875, one of those seedling trees became known in Wisconsin for the tremendous size of its fruit.

What's interesting about Wolf River is what it doesn't have. Insipid flavor and mealy texture make it downright lousy for fresh eating, and it doesn't keep that well, either. So why is this apple still in existence? Well, it's immense, for one thing, and quite attractive. It makes a powerful statement, this huge red apple with the emphatic crown of russet. And its very dryness makes it one of the best baking apples, where that dry flesh magically transforms into a caramelized, creamy delight. It is also a great drying apple, should you feel a sudden burning desire for dried apples. The large size yields nice, big, chewy rings in just a few days. Fedco Trees's John Bunker—who once had to set straight a woman who believed that as a child she had eaten "Wool Fibbers" in her grandparents' yard—favors Wolf River in pie. In *The Botany of Desire*, Michael Pollan singled it out for disdain: "It had the yellow, wet-sawdust flesh of a particularly tired Red Delicious without even a glint of that apple's beauty." No argument on the flavor—no one ever intended the Wolf River for fresh eating—but I have seen gigantic Wolf Rivers that almost vibrate with robust apple beauty.

York Imperial

Alias York, Johnson's Fine Winter **Origin** York, Pennsylvania, 1820. **Appearance** Large and lopsided, sometimes comically so. Often cannot stand up on its own. Heavy in your hand. Classically striped carmine and salmon, in the tradition of Northern Spy, with raised dark dots of russet. The stripes turn dull red in storage. **Flavor** Fresh off the tree in September, it has a classic sweet-tart apple flavor and a noticeable astringency. Not intense, but very refreshing. After a couple of months of storage, it turns sweet and rich, with a strong Juicy Fruit gum flavor, and just enough acid. **Texture** The dreadful skin should be peeled. Inside, the very juicy flesh does not snap, but neither does it mush. Firm and substantial. **Season** Best in winter. **Use** A leading sauce apple, superb keeper, excellent for drying. **Region** Hugs the Mason-Dixon Line: Pennsylvania, Maryland, West Virginia, Virginia.

In the early 1800s, a farmer named William Johnson lived near York, Pennsylvania. Infirm and housebound, Johnson would pass the days in a chair at his window, watching people pass by on the road outside his house. He began to notice that each spring, the local schoolchildren would dig through the leaf litter beneath a particular seedling tree beside the road, pull out large red apples, and devour them. Johnson was impressed that any apple could survive beneath the snow all winter, so he eventually attained a few of the apples and found them to be firm and delicious. He brought the tree to the attention of the York orchardist and watchmaker Jonathan Jessop, who was also impressed with the apple. Jessup grafted it and in 1820 began selling young trees of Johnson's Fine Winter, touting its keeping qualities. Nobody was interested, so in disgust Jessop dumped the unsalable trees in a hollow near his place. Well, that got the attention of the frugal farmers of York County, who eagerly salvaged the free trees. Eventually, word got around about the York apples with the extraordinary staying powers, and around 1850 a well-known pomologist dubbed it "the imperial of keepers" and changed the name to York Imperial.

By the turn of the century York was the leading apple of the region, a direct competitor for the English export market with Ben Davis, which was grown throughout the Midwest. York lasted as long as Ben Davis and tasted way better. But in 1930 England imposed import restrictions on apples, killing that market and triggering the rise of a local processing industry to take advantage of the excess fruit—an industry that still survives. If you enjoy Musselman's Apple Sauce, chances are good you are eating York Imperials. Many dried apples are also made with the dense fruit.

Though never seen elsewhere, York Imperial continues to be a regional favorite in the Mid-Atlantic, and is still (barely) one of the top fifteen commercial varieties in the United States. Every Thanksgiving, I see Yorks piled high in Maryland and Virginia markets. They are easy to spot, because they look like they are posing for a Salvador Dalí painting, so off-kilter that they appear to be sliding sideways.

Zabergau Reinette

Origin Zaber River, Germany, 1885. **Appearance** A beautiful russet apple with golden yellow skin glowing through the coppery coat. Sometimes a few tiny streaks of red show up. Not unlike St. Edmund's Russet, or even a giant Chestnut Crab. **Flavor** As mouthfilling as a breakfast cereal, Zabergau Reinette is all about sweet and nutty, with a toasted almond–cookie crunch and a squirt of lemon. **Texture** Crunchy and dense. You have to really shear off pieces. Not especially juicy, but not dry. The thin russet skin is unnoticeable as you power through. **Season** Pick in October. Let it sit for a month or two, unless you are feeling particularly masochistic. **Use** Great all-purpose apple. Eat fresh or bake in crisps. Fry in bacon fat. **Region** Germany is its stronghold, but it is now on a blitz through the hip orchards of America.

Like Riesling, its fellow German fruit, Zabergau sports a honeyed sweetness shot through with gum-wrenching acidity that is hard to forget. It's the Teutonic counterpart to Ashmead's Kernel, and, in a good year, one of the world's great tasting apples. Other years, its flavor can be surprisingly watery. Like many German varieties, it was not "discovered" by America until fairly recently, but now all the coolest apple people have one.

KEEPERS

For hundreds of years, one of the primary appeals of the apple was its tenacity. Unlike most fruits, apples, when kept in the cool moistness of a root cellar, could last for months, providing a welcome dose of freshness in the darkest days of winter or early spring, when even strawberries were still months away. Not all apples, mind you; many apples disintegrate within a month of being picked. But over the years, certain "keepers" with extraordinary longevity were discovered and propagated. These keepers are easy to identify because, straight off the tree, they are about as palatable as a baseball. But that cellular hardness is what makes them last; they need a few months to soften up, and they can go for a few months beyond that. Combined with summer apples, a selection of mid- and long-term keepers ensured that the well-organized farm family could be polishing off the last of the Arkansas Blacks in June, just as the first Yellow Transparents were arriving, and would never be without fresh apples.

Today, technological innovation has reduced keepers to historical footnotes. Beginning in the 1960s, apple growers learned that by storing apples in controlled atmosphere (CA) warehouses, they could arrest the breakdown process. Apples require oxygen to ripen—and to over-ripen. In CA storage, apples are kept in rooms filled with nitrogen gas, at temperatures just above freezing and humidity levels of around 95 percent. With virtually no oxygen, they will stay in a state of suspended animation for up to a year and still be marketable. (Be warned, however, that, as in *The Picture of Dorian Gray*, their souls are still aging, despite appearances to the contrary; once released from CA, they tend to go downhill fast, and can even sabotage you in recipes.)

Not even the hardiest Arkansas Black will have the crispness of a fresh-picked apple after six months in storage. Yet I think keepers can still enrich our lives in a way that CA Galas and New Zealand Braeburns can't. Just as the complex flavor of prosciutto develops only after months of aging, as the flavor precursors in the fresh meat undergo a variety of chemical reactions to become the screamingly delicious flavor compounds we know and love, so the flavor of an "aged" Arkansas Black is something no brand-new apple will ever have. Those flavors, once eagerly anticipated by generations of farm children, have disappeared from our modern table, but, as more of the classic keepers make a comeback, they are once again finding their way into our lives.

If you'd like to store your own apples, it helps to have a root cellar or a room that mimics those conditions: temperature no higher than 40 degrees, humidity no lower than 50 percent. Your refrigerator can also work pretty well if you store the apples in a plastic bag with a few holes punched in it to allow the ethylene gas produced by the apples (which is what triggers the ripening process) to escape, while maintaining a relatively high humidity. Most important, choose apples with hard flesh and a naturally waxy skin, which minimizes moisture loss (and the mealiness that ensues).

Ananas Reinette

Origin Netherlands, early 1800s. **Appearance** A small, yellow apple with green dots just beneath the surface and some raised, triangular brown lenticels. With time and sun exposure, the skin blushes the yellow-orange color of a Meyer lemon. There is often a touch of russeting around the stem. **Flavor** An acid bomb. There is sugar there, and aroma, but it's hard to fathom through the blast of fresh-squeezed lemon juice. With full tree ripening, the famous pineapple flavor arrives. **Texture** The skin is so stiff you can feel its firm knife edge with your finger. It will cut your gums as you chomp into it, and then the acidic juice will give you a thrill. The flesh is very firm, but also quite fine grained. **Season** November; will last until spring. **Use** Fresh eating once fully ripe; excellent flavor for sauce and pies, though the small size can be frustrating. **Region** Rare everywhere (except maybe Germany), but grown by heirloom enthusiasts from Vermont to California.

These unique looking apples must have some of the most outré genetics of any of their kind. They don't really look like apples at all; they remind me of certain guavas. *Ananas* means "pineapple," but the pineapple flavor really doesn't emerge until the acid diminishes. Ananas Reinette was popularized in Germany in the nineteenth century, and never really took off in the United States, but it's a perfect apple for places like California, where there is enough heat to really build the pineapple flavor.

There's much controversy as to what the French term *reinette*, which has been applied to many apples, means. (See the glossary for a complete discussion.) One theory is that it is a corruption of *rainet*, which means "little frog." I'm skeptical, because most of the reinette apples look nothing alike, but Ananas Reinette is the one reinette that, if you squint hard, does look a bit like a tree frog—a pineapple tree frog, I suppose.

Arkansas Black

Origin Bentonville, Arkansas, 1870. (Winesap seedling.) **Appearance** There's something nice and reassuring about holding an Arkansas Black. In the hand, it feels like somebody carved an apple from a piece of oak and painted it a blackish burgundy. It's that hard. Sometimes it's a uniform deep purple, sometimes a russeted dark red. It's small, round, smooth and dry skinned. The white flesh stays greenish until winter, when it turns a creamy yellow-orange. **Flavor** Surprisingly delicious. Although powerfully astringent at first, after a few months of storage, Arkansas Black becomes bright and perfumed, lightly sweet, and juicy. It has a flowery, tannic liveliness to it, like a glass of iced tea lightly sweetened with orange-blossom honey. Later in winter, as the acidity fades, an unmistakable honeydew melon flavor emerges. **Texture** In fall, the apple is bulletproof. Wait until winter, for beneath that steely exterior is a fine grained and juicy delight with good snap. **Season** Pick in October, enjoy December to April. Once it decides to get mealy and greasy, though, it doesn't hold back. **Use** A great frying apple. Excellent for all culinary uses in late fall and early winter, and good for fresh eating in late winter. **Region** The South, particularly Arkansas and the Ozarks. Probably as popular now as it ever was.

In this age of instant gratification and air-shipped produce, we have forgotten the whole category of apples that were designed for storage, which eventually get as tender and sweet as typical dessert apples, and have their own unique suite of flavors. The foremost example of a long-view apple is the Arkansas Black, which originated as a seedling from a Winesap apple in Benton County, Arkansas, in 1870. It must have stood out as soon as it fruited—nothing else in the apple universe looks like an Arkansas Black, which comes off the tree red-purple and just gets darker through the winter. Its handsome looks have distracted from its other fine qualities—it is one delicious apple. True, it has impermeable skin, but the nice thing about tough skin is that it peels so easily—and this is definitely an apple to peel.

It didn't take long for the good people of Arkansas to figure out that they had something special on their hands. By 1907, as much as 15 percent of all the apples grown in Arkansas were Arkansas Blacks, which were so hard and late-ripening that they could be ignored until all the earlier, more fragile apples had been dealt with, and could then survive both shipping to distant markets and storage in the icehouse until spring. It didn't hurt that they were gorgeous. At that time, Benton and Washington Counties had more than 4 million apple trees—more than any other counties in the country. But by the 1930s, the drought and the depression it helped trigger wiped out the Arkansas apple industry, which never returned. Arkansas Blacks, however, remained popular with hobbyists throughout the region, where today they provide an excellent taste of the bygone South.

Ashmead's Kernel

Origin Gloucester, England, early 1700s. **Appearance** Small to midsize, often warty, with a simple honey-green russet coat and one orangey cheek. The quartz flesh is generally laced with green. **Flavor** Like a tart homemade lemonade, Ashmead's Kernel pairs intense sweetness with gum-searing acidity. **Texture** Very hard and fairly dry. **Season** Pick in October, but don't eat until Christmas, unless pain is your thing. **Use** Fresh eating, pies, hard cider. Keeps well. **Region** Found in most serious apple collections. Sold commercially by many heirloom specialists.

"What an apple," the English gourmet Morton Shand said of Ashmead's Kernel back in 1944, "what suavity of aroma. Its initial Madeira-like mellowness of flavour overlies a deeper honeyed nuttiness, crisply sweet not sugar sweet, but the succulence of a well-devilled marrow bone. Surely no apple of greater distinction or more perfect balance can ever have been raised anywhere on earth." Long a cult favorite, Ashmead's Kernel will surprise you (as, indeed, three hundred years ago it must have surprised William Ashmead, the lawyer who raised this "kernel"—pippin, in other words—in his Gloucester garden). You wouldn't look twice at the homely little apple on the street, but beneath the drab clothing lies a wild child that will push the extremes of honeyed sweetness and racy acidity until you cry for mercy. "It's a delicious trip to that fine line between pleasure and pain," says Steve Wood of Farnum Hill Ciders. He uses Ashmead's Kernel in many Farnum Hill blends to add "mad florals"—mango and guava aromatics— and especially to instill that knifelike acidity that makes your mouth water for a second glass, or a second bottle. Every few years, a batch of Ashmead's is so good that Wood bottles it unblended as a Muscadet-like high-acid, high-alcohol cider.

Ashmead's Kernel was one of the many English varieties of apple to come to the New World early in the 1700s, but unlike most Old World varieties, it thrived in the extreme North American climate and has maintained a place in orchardists' collections ever since. Many a tart apple fan, upon tiring of tarrying with Granny Smith, has found a new zest for life after a walk on the wild side with Ashmead's, which is now in the midst of a modest New England revival.

Ben Davis

Aliases New York Pippin, Funkhouse, Funk Apple, and many more **Origin** Berry's Lick, Kentucky, circa 1800 (farm of Captain Ben Davis). **Appearance** A large, classically shaped apple, wide at the shoulders and tapering toward the foot. In early fall, Ben Davis starts out green with a wash of carmine on one side and russet spilling out of the stem, then turns brilliant red all over with darker stripes and streaks of gold peeking through. **Flavor** A Ben Davis grown under less-than-perfect conditions has all the charm of a croquet ball. A southern-grown Ben Davis is not as bad as advertised. In early fall, it even has a refreshing tartness, with an odd mint or tarragon note. By midwinter, it's a bit cottony, but perfectly sweet and edible. **Texture** Dry, but crisp enough. In the hand it feels very hard yet strangely light, a function of the air between its cells. The skin seems as though it were designed for interstellar voyages. **Season** Pick in October. Keeps all winter. **Use** Excellent for selling to people who have never tried one before. A passable baseball stand-in. Good still-life model. **Region** Centered around Arkansas, Kentucky, and Tennessee, but blessedly rare even there.

Those of us who shake our heads at the sad state of industrial agriculture tend to have a romantic image of the Golden Age of Ag, when people grew things for taste, rather than mercenary qualities like portability and shelf life. These romantic images are easily shattered by the rock that is Ben Davis. From the Civil War until the early 1900s, it was the dominant apple of the South and may well have surpassed Baldwin as the most-grown apple in America. Baldwin held the North, but Ben Davis swept the South. (Northerners tried their luck with Ben Davis, too, but found that it was virtually inedible when grown in cold regions—though this didn't necessarily dissuade them from cultivating it.)

In the late nineteenth century, the largest commercial orchards yet seen in America sprang up throughout the South, with Arkansas and Ben Davis leading the way. To this day, it is one of the easiest apples to grow, setting large crops of big, cosmetically perfect, Methuselah apples every year. It was known as Mortgage Lifter for the profits it brought its growers, who sent barge after barge of Ben Davis down the Mississippi to the Port of New Orleans, and on to Europe, where they underwhelmed an entire continent with their bruise-free exterior and cottony interior. As railroads penetrated southern farm country, Ben Davis turned up in groceries throughout America, too, where it could last until the following summer.

Despite its market success, Ben Davis quickly developed a reputation. Lee Calhoun, the southern apple historian, quotes an account from 1911 that describes the apple as "very popular with hotel keepers. Few patrons have the hardihood to bite into one. Occasionally a greenhorn will bite into a Ben Davis." He also describes how the particularly wooden northern variety of Ben Davis was shoveled into freight cars like so much coal. The Maine pomologist John Bunker told me an old joke about Ben Davis, which goes something like this: There once was a college agricultural professor who bragged of his ability to identify any apple by taste alone, so one day his students decided to put him to the test. They blindfolded him and laid out a variety of different apples before him. Last in line was a cork they had dipped in Ben Davis juice. The professor tasted down the line, correctly identifying each apple with ease. When he got to the cork, he bit, chewed, swallowed, and looked thoughtful for several moments. Finally he said, "Well, boys, you've got me stumped. That tastes like a Ben Davis, but I'll be darned if it isn't the juiciest Ben Davis I ever had." This is why Ben Davis is sometimes called the Red Delicious of the nineteenth century. By World War I, it had been usurped by Red Delicious itself, which, ironically, originally came to fame as the tastier red apple, before decades of selecting for color over flavor turned it into the Ben Davis of the twentieth century.

Don't confuse Ben Davis with Black Ben Davis, an apple that has always had a Faulknerian ring to me. ("There was a man and an apple too this time. Two apples, including Black Ben Davis, in whom some of the same blood ran which ran in Arkansas Black...") Black Ben Davis arose from the compost pile of an Arkansas apple-drying operation around 1880 and was named for its resemblance to Ben Davis, though it was unrelated. It's a very good apple, firm and juicy with a scrumptious balance of sweet and tart, and with lots of strawberry-pear aromatics. Was it the best apple in the world? No. Could it go toe to toe with a Cortland? Yes.

Bethel

Origin Bethel, Vermont, 1800s. **Appearance** Gorgeous, with brick-red streamers of color and white sparkles cascading down its sides. Late in the fall it becomes a uniform dark red with purple streaks. **Flavor** Mild all around. Neither terribly sweet nor sharp nor aromatic. The skin delivers a vegetal, green-pepper finish that is not entirely pleasant. **Texture** Firm, but not crisp. Dense. **Season** A winter apple, it can stay on the tree until November and will last for several months. **Use** A decent baker. **Region** Northern New England and Canadian borderlands; uncommon.

Bethel is semiofficially lumped in the "Blue Pearmain Group," and indeed, it could be Blue Pearmain's little brother, the less accomplished one that has slunk off to live in Vermont. It has something of Blue Pearmain's melon fruitiness, without the tropical intrigue. As you might expect from Bethel's Vermont beginnings, this striking apple's main claim to fame is its winter hardiness. It keeps well, but doesn't excel at anything in particular and at this point is simply holding on in a handful of spots along the northern tier of New York and New England, cozying up to boreal, self-reliant types, waiting for the grid to go down and its fortunes to rise again.

Black Oxford

Origin Paris, Oxford County, Maine, 1790. **Appearance** Green-skinned and splashed with brick-purple in early fall, it transforms into an extraordinary purple-black by November. At first the skin is dry with raised, papery, greenish white stars, but as the purple comes out it turns bright and waxy. No apple feels so much like you're gazing into the sky on a clear, icy New England night. The long-lived tree has markedly pink blossoms in spring, and looks like it's filled with plums in late fall. **Flavor** In fall, it is quite sweet and rich, like a buttery vanilla applesauce, but still green inside, with an astringent aftertaste of cattail, or bamboo shoot crossed with spinach. By December, the grassiness fades and a complex and exhilarating mix of sweet, tart, and tropical builds for the next month or so. **Texture** Black Oxford is sometimes known as "the rock." Beneath that purple armor, the flesh tenderizes by the winter solstice. **Season** Leave it on the tree until early November. Start tasting around Thanksgiving. **Use** A supremely talented apple that can do most anything asked of it. Peel and bake it in the fall; eat it fresh Thanksgiving to Easter (though it'll keep until May Day); or make a pink sauce with it anytime (that tough skin dissolves surprisingly willingly). **Region** Downright common in Maine, once again; rare elsewhere.

Maine's greatest apple and an A-list celebrity in the new world of apples, Black Oxford was virtually forgotten back in 1979, when John Bunker (later of Fedco Trees) was managing the tiny co-op in Belfast, Maine. One fall day, an old farmer named Ira Proctor walked in with a bushel of black apples he hoped to sell. Bunker bought them all. That was the beginning of his obsession with the Black Oxford, aged specimens of which he continued to find here and there in the Maine countryside. Bunker adored its deep-purple skin, rich flavor, versatility, and epic keeping qualities. It became his favorite apple and a Fedco centerpiece. "Over the years we have grafted thousands of them for our Fedco Trees catalogs," he writes in his book, *Not Far from the Tree*. "We are repopulating the earth with Black Oxfords."

Black Twig

Alias Mammoth Blacktwig, Arkansas **Origin** Washington County, Arkansas, 1842. (Winesap seedling.) **Appearance** A round, solid red or red-over-yellow apple, with whitish dots cascading toward the eye, slightly waxy skin, and a splash of russet in the stem bowl. Looks a bit like a southern Baldwin. Solid and substantial, this is an apple's apple. **Flavor** Starts off tart, with a touch of Winesap's tannic quality. The sweetness and acidity come into beautiful balance around December. **Texture** Has the unyielding skin you'd expect in a keeper. In fall, it feels like a block of wood. By late winter, the flesh feels coarse and a little soft. **Season** Pick in October. Keeps for months. **Use** Makes terrific, firm, tart pies in fall. Good for hard cider and as a root-cellar keeper. Becomes quite delicious around Christmas. **Region** Throughout the South, especially the western side of the Appalachians.

In 1842, when Winesap already ruled the South, a settler named John Crawford arrived in Washington County, Arkansas, and planted an orchard from Winesap seeds. At the time, Winesap was often known as Blacktwig, so when one of Crawford's seedling trees grew up and began producing apples with a distinctly Winesap-like appearance and flavor, but a size 50 percent greater, he named it Mammoth Blacktwig. The apple remained obscure until it was entered in the 1884 New Orleans World Expo, where it got renamed Arkansas. The apple was a hit, but much naming confusion followed, for there was another Winesap seedling (now known as Paragon) going by the name of Big Blacktwig. Many declared the two to be the same apple—confusion that persists to this day. Eventually, the other seedling stopped being called Big Blacktwig, so Mammoth Blacktwig was often shortened to plain old Black Twig. Today, you'll see it both ways. However you see it, don't let anyone tell you that it's a Paragon, or that it originated in Fayetteville, Tennessee. Different apple.

Black Twig had a great career among Appalachian home orchards, where its Winesap flavor and keeping qualities, heft, disease resistance, and preference for poor soils were much appreciated. Commercial orchards stuck with Winesap and its more robust output, but Black Twig remains a favorite with apple people throughout the South.

Court Pendu Plat

Aliases Court Pendu Rouge, Wise Apple **Origin** Known in Normandy in the 1500s, but possibly considerably older. **Appearance** Looks like a Cox's Orange Pippin that got stepped on. Doughnut shaped, with antique orange-red skin, green underneath, and some gentle russeting. **Flavor** Mysterious and unfamiliar, with hints of ancient kitchen spices, along with pineapple and pear. **Texture** Solid in the hand. Crunchy. The density makes for a full-jaw workout, and frankly it can be a bit exhausting. Virtually juiceless. **Season** Pick in October, try in December, keep until spring. **Use** A unique fresh eating experience; an inspired, surprising choice for tarts or baked apples; and probably a deeply traditional apple for mincemeat pies. **Region** Nearly extinct in France, the United Kingdom, and the United States.

This is one of the oldest apples in Europe and, to my mind, the most medieval. Pluck one from a small tree, bite into it, and you can channel the Cistercian monks, preserving these mysterious gifts from God within the secrecy of the abbey walls. "Court Pendu" refers to the apple's short stem; on the tree, it looks as though the apples, which grow in pairs, are attached directly to the branches. Somehow this makes them seem even more medieval, like a living block print. "Plat" refers to their flat shape. Court Pendu Plat had its heyday in 1600s France and Elizabethan England, where it came to be called Wise Apple for its habit of blooming very late (after the last frosts had passed), and it has been staying about half a step ahead of oblivion for the past few hundred years.

People tend to lump the words "crisp" and crunchy" together when describing apples, but Court Pendu Plat makes the distinction clear. It is very crunchy, but not at all crisp. It doesn't break apart in pieces; it forces you to wrestle off every scrap. Weird, but weirdly addictive once you dig into it. Just don't plan on doing anything else at the same time.

D'Arcy Spice

Alias Baddow Pippin **Origin** Tolleshunt D'Arcy, Essex, England, 1785. **Appearance** Small, splotchy, greenish, russeted apple with a milk-paint carmine blush on one shoulder, pink freckles, and a splash of faded orange around the stem. **Flavor** An intensely yummy mouthful of acid and sugar and that famous spice note, which some describe as nutmeg, but to me is more allspice. There is lots of blood orange in the mix. **Texture** Hard, crisp, and snappy, like a raw sunchoke. The interior breaks off in marbled chunks like a block of aged Parmesan cheese. The thick, russet skin is unpleasantly chewy. **Season** Traditionally, in Essex, people waited to pick D'Arcy Spice until Guy Fawkes Day (November 5). Let them sit for a month after picking for the full "spice" to develop. They will last for a few months. **Use** A very late keeper for eating fresh or cooking. In early winter, once Cox's Orange Pippin has passed, D'Arcy Spice picks up the mantle of best-tasting thing around. **Region** Outside of Essex, England, these can be tough to find, though they have some huge fans in the heirloom apple community in the United States.

This uniquely spicy and high-flavored apple was spotted in 1785 in the gardens of D'Arcy Hall, the fifteenth-century manor of the founding family of Tolleshunt D'Arcy, a village near England's eastern coast, and became celebrated in nineteenth-century England. To this day, the approach to D'Arcy Hall is lined with D'Arcy Spice apple trees. Zeke Goodband, of Vermont's Scott Farm, says it's the only apple he ever takes home with him (partly because of its high flavor, and partly because it's the last apple of the year to be picked). Although hard to find, this one should go high on your bucket list.

GoldRush

Origin Purdue University Horticulture Research Farm, West Lafayette, Indiana, 1973. (Seedling of Golden Delicious.) **Appearance** Clearly a homely child of Golden Delicious, GoldRush's spiderweb russeting, color, and texture remind me of a sick leaf. Once fully ripe, however, it turns a beautiful, freckled gold. **Flavor** Tart and rhubarblike in September, turning intensely sweet and spicy, yet still with that sour, metallic counterpunch, in November. **Texture** Very, very hard, yet very juicy. An unusual combo. **Season** Usually picked in November, GoldRush is a champion storer. Look for it all winter. **Use** Delicious fresh, excellent in salads because it doesn't brown, and perky in pies. **Region** Many people in the Mid-Atlantic and Blue Ridge region tell me GoldRush is now their favorite apple. Less known elsewhere, but that may well change.

GoldRush feels more rigid than other apples, as if it had better cellular scaffolding, which apparently it does; it earned one of the highest scores on a pressure test of any known apple. This is what is responsible for its extraordinary crunch and longevity. It was bred by Purdue University to be resistant to scab and other apple diseases (the resistance stems from a crab apple that was its great-great-great-grandparent), was released to growers in 1994, and is becoming quite popular among organic and low-spray apple growers.

Keepsake

Origin University of Minnesota, 1979. (Cross of Northern Spy and Malinda.) **Appearance** Has an unmistakable mottled look. The smooth, waxy, greenish white skin is covered in red blotches, rather than stripes. The red blotches bleed into one another, like camouflage on some fantastical beast. I actually find it quite attractive and ancient looking; others think it's downright ugly. The flesh has a greenish tinge unless dead ripe. **Flavor** Very sweet with nutty, flowery notes, like pecans simmered in rosewater, with a touch of lemon. Great for those who like sweet, flowery apples. **Texture** Hard as a rock, but it breaks instantly with a sharp snap, calving like an iceberg and spraying juice. The fine-grained flesh is smooth and creamy, but the skin is chewy. **Season** Ready in October, but will last well into the winter. **Use** Best for fresh eating. **Region** Has a decent foothold in upscale markets and orchards across the northern states.

There's no doubt that the Honeycrisp got its explosiveness from its mommy; what's perplexing is why the mommy isn't better known. Keepsake, which was developed by the University of Minnesota in the 1960s and released in 1979, has nearly the aural pyrotechnics of its famous child, along with much better flavor. As you'd guess from the name, it's also a first-rate keeper. A great apple all around, though little known. Might its age-spotty appearance be holding it back? A sleeper.

Lady Williams

Origin Paynedale, Donnybrook, Western Australia, 1935. **Appearance** A small, tight, green apple turning candy-apple red late in the season. Smooth skinned. As you can see from the faint auburn blush in the photo, these were just beginning to turn their thoughts to ripeness; the flesh was quite green inside. A few more months and they would turn red on all but the shady side. **Flavor** Always tart, Lady Williams balances its gooseberry zip with loads of sugar if left on the tree into winter. **Texture** Firm and crisp, this apple rarely softens. It rarely gets juicy, either, and the skin is not insubstantial—all qualities you expect in a champion keeper. **Season** Incredibly late. In California, they sometimes don't pick Lady Williams until February. In northern climates, it will never ripen. It will store until summer. **Use** Good culinary apple. Loves the root cellar. Eat fresh…eventually. **Region** Possibly the best apple for hot southern climates. In the United States, it has a real presence only in Southern California, but it has been quite popular in Western Australia since the 1970s.

No other apple in the United States behaves quite like the Lady Williams, which needs hot, long growing seasons to ripen properly. Only then does the acid become balanced with an intense sweetness. This means it fails miserably in most apple-growing regions, but succeeds where almost all other apples fail.

Lady Williams had a rough start in life. The original seedling was born in 1935 on Bonomia Farm in Western Australia, but was accidentally cut down a few years later. The tree came back, and the farm owner, A. R. Williams, harvested the first apples from it in the 1940s. He named it for his wife, who went by Lady Williams because her first name was apparently unpronounceable, and its unparalleled keeping qualities brought it fame and fortune Down Under. A later tryst with Golden Delicious produced Pink Lady (aka Cripps Pink). Lady Williams has never truly caught on in North America, though it is a natural for the Southwest and Mexico.

Malinda

Origin Orange County, Vermont, early 1800s. **Appearance** Medium-size, light-green apple with a burst of russet emerging from the stem and a slight salmon blush on one cheek. **Flavor** Sweet, tropical, and mysterious, with a hint of Caribbean exoticism, papaya, or perhaps creamy, custardy soursop. No sign of acid. **Texture** Crunchy and dry. **Season** Pick after a hard frost, after which it will keep in the root cellar until spring. **Use** Supreme keeper and a hardy northern tree. Makes excellent dried apples. **Region** Found primarily in Minnesota, and occasionally other states along the Canadian border.

Malinda was a semipopular apple in its native Vermont, and in Minnesota after the 1860s, based on the tree's ability to thrive in the cold and on the apple's ability to stay crunchy in the root cellar until the dandelions bloomed. But its one-dimensional taste (all sweet, no sour) and its cosmetically challenged appearance always made it a second-string apple. Yet the breeding wizards at the University of Minnesota always saw something in Malinda they liked, and they have leaned heavily on Malinda for their most dazzling inventions. Malinda is the parent of Chestnut Crab and Haralson, the grandparent of Keepsake and Sweet Sixteen, the great-grandparent of Honeycrisp, and the great-great-grandparent of SweeTango. That's an extraordinary line, with all apples sharing some of that explosive "apple Cheeto" texture that is the essence of the modern apple. So while Malinda couldn't be more old-fashioned herself, she can be considered the Mother of Modernist Apples, the way that Cezanne was the Father of Modernist Painting.

Roxbury Russet

Aliases Roxbury, Boston Russet **Origin** Roxbury, Massachusetts, early 1600s. **Appearance** The classic russet, green skin turning the color of oiled pine in the sun, covered in a sandpapery russet. Let's face it: it's an ugly apple, although it will sometimes develop a brassy glow on one cheek. Tends toward roundness, though sometimes oval and sometimes squarish. It seems to have a lot of variation. **Flavor** Yummy and strange. Early in the season, its sweetness is almost completely overrun by aggressive, almost painful acid. It's like biting into a kumquat. By early winter, the acid gives way to a delicious, rich persimmon with nutty undertones. **Texture** Very hard and crunchy at first, Roxbury Russet can turn spongy if not properly stored. Its dense, white flesh is granular and dry, and its skin, like most russets, is thick but not tough. **Season** Pick in October. Gets tastier for several months if stored properly. **Use** Excellent in pies and crisps. Superb keeper. (Store in plastic to preserve moisture.) I love it fresh. **Region** Still highly esteemed throughout New England.

This is the oldest American apple, having sprung up on a hill above the Massachusetts Bay Colony, and already popular throughout southern New England in the 1600s. It had conquered the rest of New England and New York by the 1700s, and had some success in the Midwest and even California in the 1800s, but its big appeal was its shelf life, and around 1900, when the technology arrived to keep many apples for months under refrigeration, Roxbury Russet became just one more russet apple that was a hard sell in the market. Yet it still has a special place in the hearts of many New England orchardists. Somebody planted one at my house about fifty years ago. In late fall, after my other apples have gone the way of all flesh, it still proudly displays a full canopy of russet nuggets, which make wonderfully nutty Thanksgiving fare.

Considering its antiquity, Roxbury Russet may well have been the variety of apple tree that ate Roger Williams, founder of Rhode Island. Williams died in 1683 and was buried on his property in an unmarked grave. In 1860, an attempt was made to exhume the body and transfer it to some classier digs, but the workmen found no sign of Roger. In his place was the root from a nearby apple tree, which had formed a perfect torso, legs, knees, and feet. The guilty root is now on display in a coffin-shaped box at the Rhode Island Historical Society in Providence.

Stark

Aliases Robinson, Yeats **Origin** Stark County, Ohio, mid-1800s. **Appearance** The color, size, and shape of a large green tomato, turning a wan magenta on the sunny side. In ideal years, Stark gets bright red. Its lenticels are quite odd: brown, raised, often triangular, and surrounded by a white halo. There is usually a touch of russet around the stem. **Flavor** Sweet, not acid, like a mild strawberry, with a vegetal finish coming from the skin. **Texture** Very crisp and crunchy, even hard. Needs time to mellow. The heavy peel holds no appeal. **Season** Pick in October (September in the South). Don't even think about touching it until January. **Use** Superb keeper. Good, if mild, baker. **Region** Nationwide, but a rarity everywhere.

This hefty apple gives a sense of substance and stability; it's not a fruit to push around. Off the tree, it can be rock hard. Stark's claim to fame is that it will last in storage until June. Back in the day, a May Stark, dug out of the root cellar and eaten while the tree was flowering with the next baby Starks, would have been most welcome. It isn't the best tasting apple in the world, but it's awfully pleasant, and in midwinter still surprisingly snappy.

Stark has always suffered from dull looks and an unmemorable name, but the tree is healthy, hardy, and uncomplaining, and the apples can withstand all sorts of rough treatment, which was enough to make this unflashy apple a commercial success throughout the country by the 1890s.

Virginia Winesap

Alias Old Virginia Winesap **Origin** Troutville, Virginia, 1922. (Seedling of Winesap.) **Appearance** Turns a very pretty carmine in early December, over a creamy yellow background, with cheerful white lenticels scattered across the surface. Nearly round, it shows touches of pink on the shoulders, like Northern Spy. **Flavor** Mildly tart and refreshing, like berry juice. **Texture** It has skin like leather. This is one to peel. The flesh is beautifully snappy and easy to eat. **Season** October. **Use** Great for cider or culinary purposes. A superb keeper. **Region** Southeast, especially Virginia.

Numerous sports and offshoots of Winesap came into being during its century-long dominion over the South. Most of these epigones were redder than the original, and that was the case with Virginia Winesap, which was promoted by Stark Bro's nursery of Missouri for a couple of decades between the wars. You still see it quite a bit in the Piedmont region. Its highest manifestation is undoubtedly Albemarle CiderWorks Old Virginia Winesap cider, one of my all-time favorites. Like Champagne with a squeeze of crab apple juice, it is tart and tangy with a mysterious bit of archaic funk, like the yeasty smell in the cupboards of your grandparents' summer home.

White Winter Pearmain

Alias White Pearmain **Origin** Indiana, mid-1800s. **Appearance** Conical, yet squat, this small apple can be green or yellow, transitioning to tequila sunrise on the sunny side. Smooth and waxy, it displays a starburst of russet around the stem. **Flavor** Sweet and perfumed, like an orange Creamsicle. **Texture** Tender and juicy, with nice smooth flesh. The skin is very thick and chewy. **Season** October in the Midwest; September in warmer regions. **Use** Excellent dessert apple. As you'd guess from the name, it's a good keeper. **Region** Excels in the Midwest, South, and California. Unlike most apples, it does not require a winter chill to produce fruit the following year.

Sometime in the 1840s or 1850s, a collection of scion wood was brought to Indiana in a saddlebag. From those grafts, two pearmain-shaped apples were grown. They came to be known as Red Winter Pearmain and White Winter Pearmain, until the red one was recognized as Esopus Spitzenberg. The identity of the white pearmain was never discovered, and so the name stuck, but the possibility lingers that White Winter Pearmain is identical to an older eastern variety, though none seems like an obvious fit—especially since White Winter Pearmain has demonstrated a knack for flourishing everywhere other than the East. It was a southern and Midwest favorite by the late 1800s. In 1870, a North Carolina horticulturalist called it "the highest flavored apple in cultivation." It's also one of Tom Burford's Top 20 Dessert Apples. Balancing that high praise is the opinion of one J. M. Hasness, Secretary of the Holt County Horticultural Society, in the 1884 *Report of the Missouri State Horticultural Society*: "Some varieties, like men, start off well, make a brilliant record for a few years, then so utterly fail as to disgust their warmest friends and admirers. Of such is the White Winter Pearmain, famous in Northwest Missouri fifteen years ago, and at that time really a fine, delicious variety, but now I pronounce it worthless." Who knows what spoiled that particular friendship. It's just one of many mysteries involving this apple, which seems to have a shady past it would like to leave behind. Another rumor has the White Winter Pearmain being the oldest of English varieties, dating to the 1200s; surely this is another case of mistaken identity for this enigmatic fruit.

Winesap

Origin New Jersey, late 1700s. **Appearance** Smallish, round, deep red, classic. **Flavor** Sweet, tart, and astringent, with a vinous twang like fresh cider that has started to go fizzy in the fridge. In storage, the acidity disappears, revealing a delightful toasted-almond flavor, like marzipan. **Texture** Rock hard and crunchy, with sharp, rigid skin. **Season** Late fall into winter. One of the best keepers. Early in the fall its flesh will be very hard and green, but it will slowly mellow and sweeten through the winter. **Use** Amazing in pies and crisps. Makes a mischievous cider with funky, herbaceous, almost saline qualities. Good dessert fruit once winter has taken the starch out of it. **Region** Winesap and its many sports are still prominent throughout the South.

Winesap is one of the heavyweights of the apple world, now fading from view, its legacy still prominent in the minds of old-timers. It is the most important southern apple of all time, the one that turned the South into an apple-growing dynamo in the 1800s. Like no other apple, it will thrive in lousy soil, of which the southern hill country had plenty, producing an annual bumper crop of tasty, red, long-keeping fruit. Growers everywhere raved about its inimitable twangy flavor—yet abandoned it for less twangy Winesap offspring (such as Arkansas Black, Black Twig, and Stayman) at the first opportunity. That twang is usually described as vinous, or winelike, and indeed the original name of the apple was Winesop, for its supposed similarity in flavor to bread dipped in wine, but to me "foxy" would be a better description, like scuppernong grapes or Darjeeling tea. That foxiness is most evident in hard cider made from Winesap, which was the apple's initial claim to fame. In 1817, the New Jersey horticultural expert William Coxe called it, "one of our best cider fruits...the cider produced from it is vinous, clear, and strong; equal to any fruit liquor of our country for bottling." Winesap's smallish dimensions—another reason it was later abandoned—didn't matter for cider.

As late as 1892, Winesap was sold by more nurseries than any other apple in America. And it was not merely a southern phenomenon. The Midwest also loved Winesap. As Washington State came to dominate world apple production in the twentieth century, it seized on Winesap as its bread-and-butter apple. Thick-skinned, tart-fleshed Winesap was a brilliant keeper, which was exactly what the growers of eastern Washington needed. You could grow the prettiest apples in the world in the high desert plains of eastern Washington, but you had no one to sell them to; there were more sheep than people in the vicinity. The markets for apples were the cities on both coasts, so growers needed an apple that could ship without bruising, and they needed an apple that kept well enough to fill that market for months. Winesap empowered Washington State to become America's apple titan. Winesaps picked in late fall would still be edible, if not prime, in June. After a brief spate of tasty early fall apples, Winesaps carried the load from winter through spring into summer, allowing them to become the number-one apple in America by World War II.

Then controlled atmosphere came along and stole their thunder. Researchers learned that oxygen and carbon dioxide were what fueled fruits' ripening process, and that if you stored them in an atmosphere of nitrogen gas, you put them into suspended animation. Virtually any apple in controlled atmosphere would keep as well as a Winesap in normal air. Growers were suddenly free to grow any apple they wanted—and they knew what apple they wanted. As Red Delicious remade the Washington State apple industry, Winesap was the first and biggest victim. It plunged from 50 percent of the apple crop in 1950 to 20 percent in 1965 and less than 2 percent in 1980. By 2002, it had disappeared from the charts.

Winter Sweet Paradise

Alias Paradise Sweet **Origin** Paradise, Pennsylvania, 1830s. **Appearance** A half-brick, half-green apple lit up by large, white, starlike dots. Looks as if it were spray painted and the paint dripped down the sides. **Flavor** Delicious guava and lychee notes are set off by a low-acid background. **Texture** Firm and crunchy. **Season** September to October, but best in early winter. **Use** With its staying power, this is clearly an apple to eat in midwinter in the snow. Sweet paradise, indeed. Known for fine apple butter and fried apples. **Region** Centered in Pennsylvania and the Mid-Atlantic states.

This lovely, shy apple makes me think of the aurora borealis, of green ribbons of cold fire swaying against the blackness. Streaks of purple or yellow luminosity sometimes paint the surface. Winter Sweet Paradise carries the cosmos within it. It stays small in the north, but can get large in warmer climes. What I describe as "guava" and "lychee," some others call "skunky."

Yellow Bellflower

Aliases Bellflower, Belle-Fleur, Yellow Bellefleur **Origin** Crosswicks, New Jersey, early 1700s or late 1600s. **Appearance** In color, form, and ribbiness, Yellow Bellflower does a commendable impression of a quince. It can be quite large and heavy, with prominent ribs showing five lobes with one peachy shoulder. The skin is very dry and unwaxed, banana-yellow, with raised brown lenticels, some quite large. The seeds in a Bellflower are large and loose, so it rattles when shaken. **Flavor** A lovely, seamless experience of sweet and tart creaminess, like a lemon meringue pie. There is the floweriness of lemon peel to go with the tang of lemon juice. The young ones have a musty finish that disappears by midwinter. Once it mellows, it becomes as rich and comforting as pudding. **Texture** On the dry side at first; leave it alone until the snow sticks. Slowly it will develop an old-fashioned apple texture, neither crisp nor soft, but dense and tender, like an underripe pear. It has a rare density for an apple, which, combined with its large size, gives it impressive heft in the hand. **Season** Comes off the tree in October tart as hell. At this point it makes great pie and alarming applesauce, but you can't eat it fresh. It doesn't really lose its starch until December, but then it makes for wonderful fresh eating right through the winter. **Use** Fresh, in tarts and lemony applesauce, winter keeper. Supposedly makes a good hard cider, because of the high sugar and acidity, but I've never encountered any. **Region** Nationwide, but uncommon everywhere.

On my list of apples that have something to teach us, Yellow Bellflower ranks high. It doesn't look like anything else in the apple world. That rumpled shape reminds me of a former beauty who has let herself go. And its flavor also represents a whole different idea of appleness. Like a Japanese dessert, it is delicately sweet and vaguely exotic. It's one of my favorite eating apples, but you have to catch it right, when the acid is perfectly moderated and the flesh is pliant but still noisy.

Yellow Bellefleur, as it was once called, has, as you might guess, particularly beautiful flowers. The tree gets huge and handsome and is long lived. By 1817, it was already well known throughout New Jersey and was "the most popular apple in the Philadelphia market," according to William Coxe's seminal *A View of the Cultivation of Fruit Trees*, which also reported that the original tree was quite old but still standing on its Crosswicks farm. The spot is inches off the New Jersey Turnpike, near Trenton, but you can imagine a glorious eighteenth-century New Jersey, with massive Yellow Bellflower trees shading its farms and filling its markets with bright yellow fruit each fall. (Today, you can still find many an ancient Yellow Bellflower shading many an old farmhouse; the fruit detective John Bunker recommends a particularly handsome one near Augusta, Maine.)

From New Jersey, the apple grew popular everywhere from Maine to the Midwest and California, where E. J. Wickson, the legendary University of California professor of horticulture (for whom the Wickson apple is named), commented on its "conspicuous excellence." But, although pretty much everybody favored the apple and every nursery catalog listed it, it was never a reliable cropper and suffered badly from apple scab and fire blight, so it never became a major player in the era of the industrial orchard. One of its offspring, however, born beneath a Yellow Bellflower tree in an Iowa orchard in 1875, did pretty well for itself: originally called the Hawkeye, it is now known far and wide as the Red Delicious.

CIDER FRUIT

There is terrific energy in many corners of the apple universe right now—thrilling new varieties being released every year, and classy heirlooms being rediscovered—but the world of cider is an absolute dynamo.

When I say "cider," I mean hard cider, the drink that kept half the English-speaking world dizzy and docile for centuries. Until the advent of pasteurization, sweet cider was a passing treat—there was no way to prevent the fermentation from starting a few days after it was pressed. Besides, why would you? Hard cider is one of the world's greatest drinks. I ferment 40 or 50 gallons of it in my basement every winter, and, though I've heard it peaks in quality about a year after being made, frankly mine never lasts that long.

Thankfully, there is plenty more to be had on the market, because America is now rediscovering the joy of cider. (And southwestern England, Normandy, and Asturias never forgot it.) After years of false starts, cider is taking America by storm, with new cideries popping up everywhere. You can hardly call yourself a serious bar if you don't have at least one cider on tap, and even the big players (such as the Boston Beer Company of Sam Adams fame, with its Angry Orchard cider, and Stella Artois with its Cidre) are getting into the mix—a sure sign that cider has returned at last.

But therein lies the problem. With cider being a brand-new drink for most Americans, we've lost the body of knowledge for understanding it. It's a bit like the domestic wine industry in the sixties and seventies, before people had learned to distinguish their Cabs and Chards and Pinots. Most people don't even know there's such a thing as cider apples, and so they buy generic bottles of "cider," which is usually pleasant and quaffable, but otherwise uninspiring. It's the equivalent of wine made from table grapes.

Real cider apples are a very different thing. Selected centuries ago for their intense acids and tannins—the bitter, puckery compounds that give good cider its lip-smacking finish—they are often inedible fresh. The fermentation vat is their reason for being, and there they create drinks that, to me, combine the best qualities of white wine, red wine, and beer, while resembling none of them. They haven't been available in the United States in great quantities, but that is changing fast, and you are about to have a front-row seat for America's second great cider revolution. You might even want to try your hand at making your own. As I said, the beauty of cider (unlike, say, beer) is that it practically makes itself. There's no brewing, no fiddling with recipes; just press (or buy) your juice, fill your carboy or bucket, top it with an air lock, stick it somewhere cool, and ignore it all winter. It'll be ready in time for some chill summer sipping.

Dabinett

Origin Middle Lambrook, Somerset, England, early 1900s. (Seedling of Chisel Jersey.) **Appearance** Small, round, speckled or striped red over a pale green background. **Flavor** Fuzzy and bittersweet, Dabinett cider is a knockout combination of peachy tannins and chanterelle fruitiness. **Texture** Who cares? If you're eating it, something's gone terribly wrong. **Season** Pick in November. Drink in summer. **Use** Hard cider. Don't eat fresh, unless sucking on tea bags is your idea of fun. **Region** Popular cider apple in Somerset, England. Now gaining popularity with cider makers in the Northeast United States.

Every autumn I fill six or seven glass carboys with juice pressed by my neighbor Terry Bradshaw, the apple guy at the University of Vermont, from blends of crazy apples. Terry does a different blend each week through the fall, usually containing anywhere from five to ten varieties, many of them wild apples or traditional English cider apples. The cider ferments in my basement and the bottles disappear throughout the spring and summer. Whenever I particularly like a cider batch, I check back to the apple list for that blend, and over the years I've noticed that the name that keeps jumping out at me is Dabinett. For all the hoopla paid to sexy cider varieties like Kingston Black and Harrison, I think Dabinett is the unsung hero of the cider world, and if I were planting a cider orchard, I'd go heavy on the Dabinett. It's a mainstay in Steve Wood's Farnum Hill Ciders (where he likes to partner it with Wickson) and in many of the classic ciders of southwest England. Sheppy's, in Somerset, makes a single-variety Dabinett, which is unusual. Generally, Dabinett needs a few friends to round it out, but when they do, it becomes the life of the party.

Ellis Bitter

Origin Newton St. Cyres, Devon, England, 1800s. **Appearance** A little, strawberry-shaped apple with yellow skin painted in brick and ginger. The skin gets quite waxy. (The weird knob on the apple in the photo is not characteristic of the variety.) **Flavor** Almost Malbec-like in its tannins, it's a full-on spitter. Some wonderful fennel notes lurk behind the bitterness. **Texture** Like many cider apples, it's on the grainy, mealy side. **Season** Pick in October. **Use** This classic English bittersweet makes formidable cider, where its sugars and tannins contribute to high alcohol and full body. Brave souls might also mix it into crisps to add a bittersweet complexity. **Region** Most often seen in its native southwest England, Ellis Bitter is making inroads among U.S. cider makers.

Any apple with "bitter" in its name is sure to be a cider fruit. This is a common apple in English cider country. Most sources say Ellis Bitter is less tannic than most bittersweet cider apples, but that hasn't been my experience. It's never been a major cider apple, but many smart cider makers use just a few percent Ellis Bitter in their blends to give a nice, dry, astringent, spicy finish.

Golden Harvey

Alias Brandy Apple **Origin** Herefordshire, England, 1600s. **Appearance** Would not turn your head on the street, or in an orchard. Golden Harvey looks like your basic, homely little yellow-green Hacky Sack. **Flavor** Tastes like an extra-tart limeade—very sweet and spicy, but it packs a wallop of lime juice that you can feel on your lips, tongue, and all the way down your throat. It's pretty addictive. **Texture** Delightfully crunchy, yet tender, with a beautiful fine grain. **Season** Pick in October. Keeps well. **Use** Although this is a brandy and hard-cider specialist, I find it quite delicious fresh, circa October. **Region** Your best bet is England's cider counties, but rare even there.

Because Golden Harvey has a lot of sugar in its dense flesh, it has long been prized for hard cider. In 1851, Dr. Robert Hogg, author of the classic *British Pomology*, called it "one of the richest and most excellent dessert apples ... it is also one of the best for cider; and from the great strength of its juice ... it has been called the Brandy Apple." Rich flavor and high alcohol is a pretty unbeatable combo, so it's unclear why this excellent apple never really took off. It's ripe for rediscovery.

Golden Russet

Origin Upstate New York, 1840s. **Appearance** A typical green-skinned russet on the shady side, but on the sunny side it turns smooth and truly golden in color, with some copper and bronze and peach mixed in. There are few sights prettier than rows of Golden Russets in the orchard, resembling an apricot sunrise. Look for a smattering of perfectly round off-white lenticels on the skin. Inside, the flesh is pure white. **Flavor** A luxurious delight of pear and fig, a ripe Golden Russet is as rich as it gets. Everything about the sweet, spicy, sprightly flavor says "apricot." When unripe, the skin has a green pepper note. **Texture** Hard and crunchy, with just enough juice to lubricate your jaw and keep it from overheating. In keeping with the theme, the texture of the skin has an apricotlike roughness. **Season** Pick in October. Don't press for cider until winter. As a dessert fruit, its flavor peaks in December and January, but it will keep until late winter. **Use** This apple does everything better than most apples do anything. **Region** Grown throughout the northern United States by people who are serious about apples and cider.

Although other apples get most of the glory, Golden Russet is the real jewel of the cider world. While the bitter-sweets and bitter sharps provide cool tannins in small amounts, Golden Russet is the workhorse, cranking out barrelfuls of full-bodied, unapologetically delicious cider, sweet or hard. John Bunker calls it "the Champagne of cider apples" and quotes a nineteenth-century authority who says, "If you plant 100 trees for cider, 99 should be Golden Russets. The other you can choose for yourself." Farnum Hill Ciders jokingly refers to it as their "Wedding White" for its ability to produce peachy, crisp, Pinot Grigio–like crowd pleasers. Many English and American ciders are made from a base of Golden Russet, spiked with more challenging varieties.

And it's not just a great cider apple. Through winter, Golden Russet stays sweet, hard, nutty, and crunchy, full of uncomplicated sugar-cookie goodness. Although quite popular in the Great Lakes region during the nineteenth century, this is one of the apples that suffered most from the red-apple fad of the twentieth century. One of Golden Russet's primary advantages was its superb keeping abilities, and once people learned cold-storage techniques that allowed them to keep red apples in decent shape through the winter, they lost interest in Golden Russet, despite its superior flavor. Now it's on my short list of heirloom apples most likely to have a second act.

Harrison

Origin Newark, New Jersey, about 1712. **Appearance** A small, pockmarked, yellow-green matte apple with tiny black specks, turning full yellow in storage. Look for the unusually long stem and the occasional pink blush. The golden-yellow flesh is sometimes laced with streaks of brown. Harrison never looks terribly healthy. **Flavor** Harrison shows off an extremely concentrated apple flavor. It's rich, viscous, and syrupy, almost like a natural apple butter. That viscous quality comes through in Harrison hard cider, which has a gingery, concentrated fruitiness and a sherrylike mouthfeel. **Texture** Dry and hard, with gritty, pearlike stone cells, Harrison is one of the densest apples I've sampled. That density, along with its fibrousness, means that your teeth sort of glance off when you try to attack. Harrison is no fun fresh, though in winter it softens up enough that you could make the effort. **Season** Late October to early November. Whether eating fresh or pressing for cider, it's best to keep it until midwinter. **Use** Makes a stunning, amber cider. Too coarse and dry for eating out of hand. Too small for baking. **Region** In the midst of a full-blown revival in western Virginia. Also on the verge of becoming the official apple tree of Newark.

In the early 1700s, Samuel Harrison planted some of the first orchards in the Newark region. Among his orchards was found a tree that yielded prodigious quantities of small, yellow-green apples that made a particularly Champagne-like cider. By the early 1800s, Newark was famous for Harrison's Sparkling Newark Cider, empty bottles of which can still be found. In 1817, the pomologist William Coxe called Harrison "the most celebrated of the cider apples of Newark" and explained that "it produces a high coloured, rich, and sweet cider of great strength," which commanded "the highest price of any fruit liquor in New York." Half a century later, in *The Cider Makers' Manual*, J. S. Buell wrote that "the most noted varieties of apples said to possess peculiar and natural properties for the manufacture of refined cider or wine, are the Harrison and Canfield, of New Jersey, from which the celebrated Newark (N.J.) cider is made. This fruit is raised upon the ridge between Newark and Elizabethtown." By then, Newark cider makers had adopted a style so similar to Champagne that their product was often called Newark Champagne and was sometimes even passed off as the pricey French import, a controversial practice that caused one critic to label it "sham-pagne."

Yet by 1900, the rise of beer had taken a bite out of the cider market, and soon Prohibition finished it off. Orchardists switched to dessert apples, the Harrison was abandoned, and the remaining trees fell to the New Jersey bulldozers.

In 1976, the Vermont fruit detective Paul Gidez began searching Newark for a living Harrison apple tree. At Nettie Ochs Cider Mill he found a single old tree. Gidez took a number of cuttings back to his Vermont orchard. Not long after that, the tree was cut down, and Gidez never found another Harrison, but from his cuttings he was eventually able to propagate 250 Harrison trees in his orchard. Recently, the Virginia pomologist Tom Burford has brought the Harrison back into circulation by singing its praises and distributing scion wood to the new crop of artisan cider makers. Burford calls it "the most enigmatic apple I've ever dealt with. When I first tasted it I had to sit down, I was so unsettled. How could it have happened that this great cider apple got pushed out of production? Anything as good as the Harrison you would think they would say, Let's take care of it."

Today Harrison is the darling of the cider world. Two Virginia cideries—Albemarle CiderWorks and Foggy Ridge Cider—have extensive plantings of Harrisons that are just beginning to bear, and a handful of nurseries are selling the trees. No one knows how Harrison will pan out—Diane Flynt of Foggy Ridge says that the flavor is superb but the growing habits are frustrating—but what an opportunity. It's as if a legendary grape called Pinot Noir had been rediscovered, and we were the first generation in more than a century to taste its wine.

Harry Masters Jersey

Alias Port Wine **Origin** Yarlington, Somerset, England, 1800s. **Appearance** Golf ball–size cider fruit, conic, cute, dark burgundy over a pale wash the color of aged cheese. **Flavor** Mildly sweet, gently astringent and bitter, with a nose of muskrat. For cider only. **Texture** Cottony dry. Inedible fresh. **Season** Pick in October to November. **Use** Hard cider. **Region** Somerset, England. Newly popular with U.S. cider makers.

A classic Somerset cider variety, still in great use. Many cider blends include a touch of Harry Masters Jersey for complexity. It has that distinctive, funky flavor called barnyard that hardcore cider people love—a blend of hay and mushroom and musk. It will not soar on its own, but can be very useful to add grip to an otherwise simple cider. Although people tend to lump all tannins together when it comes to apples, there are soft tannins and hard tannins, and Harry Masters Jersey is an excellent source of the soft, more palatable kind, which give a feeling on the tongue similar to soft red wine, hence the apple's other name. Harry Masters was a horticulturalist in Yarlington Mill who propagated some of the most famous cider apples, including, yes, Yarlington Mill and Harry Masters Jersey.

Hewes Crab

Aliases Virginia Crab, Virginia Hewes Crab, Hewe's Crab, Hugh's Crab, Hughes Crab **Origin** Virginia, around 1700. **Appearance** So pretty. Bright pink and yellow little Christmas ornaments. A tree full of these will smack the crabbiness right out of you. **Flavor** Makes a yeasty, biscuity hard cider, like fine Champagne. Sour, astringent, and virtually inedible fresh. **Texture** Nasty, woody flesh. **Season** Press the fruit in October; drink the cider year-round. **Use** Cider, plain and simple. **Region** Virginia (but will thrive as far north as Canada).

Hewes Crab's identity is tied strongly to colonial Virginia, but we don't know exactly where. William Coxe, America's first apple authority, was evasive in 1817, saying only that he knew of century-old trees "now standing in the orchard of a respectable inhabitant of that State, from whom I obtained the information." This looks to be one of the oldest American apples, and possibly the first Virginia native to take off—purely for cider purposes. Coxe hit it on the head when he wrote, "The juice, although acid and austere to the taste when mixed with the flesh, becomes sweet and highly flavored when expressed from the pulp in the perfect maturity of the fruit. The flesh is singularly fibrous and astringent; in pressing, it separates from the liquor, which runs through the finest flannel like spring water."

In other words, unlike most apples, in which a lot of brown solids dissolve with the juice when pressed, the Hewes Crab retains its solids in that fibrous mesh and releases pure, light juice ready to ferment into a super-clean cider with none of the funky qualities common to so many ciders. For this reason, it was the centerpiece of Thomas Jefferson's cider dreams at Monticello. Washington also preferred "crab cider" to all others. A classic 1848 book on American fruits put it thus: "The Virginia Crab makes a very high flavoured dry cider, which, by connoisseurs, is thought unsurpassed in flavour by any other."

Of course, the same qualities that favored Hewes Crab for cider made it an utterly useless apple for the past century or so. Bite into a Hewes Crab and you'll be left with a mouthful of lime and quinine; it's a brutal gin tonic of a fruit. The bitter, astringent finish carries fruity, hopslike perfume notes. Somehow that all sugars out into a stunning cider, demonstrating the law that you can't really predict the quality of the cider from the taste of the fruit.

If you haven't encountered a Hewes Crab cider yet, you will soon. It is once again at the center of Virginia's cider revival, with Albemarle CiderWorks, Foggy Ridge Cider, and Castle Hill Cider all featuring it, and the trees are once again thriving at Monticello. You'd be hard-pressed to find a finer single-varietal cider than Albemarle CiderWorks's Virginia Hewes Crab reserve, which has a brioche nose to rival Champagne and a robust (for cider) 10 percent alcohol. (There are, broadly, two styles of Champagne—a crisp, light-bodied style often described as "appley," and a richer, full-bodied style often compared to yeast, biscuits, and brioche—and ciders can resemble either. Hewes Crab makes a biscuity cider, while Newtown Pippin makes a lean, appley one. Lean is fine, but go for the biscuity Hewes if you can find it.) Even New York's Hudson River Valley is getting in on the game, with Annandale Cidery producing a superb Hewes Crab cider, the first I've seen outside of Virginia.

Kingston Black

Origin Kingston St. Mary, Somerset, England, circa 1820. **Appearance** Kingston Black looks distressed. Its deep mahogany skin could use a new coat of paint. A beige russeting pours out the top and down its shoulders like the hair on an aging rock star. It leans off-kilter and has trouble staying upright. **Flavor** The white flesh is surprisingly plain, just semisweet and unmemorable. The magic is in the dark skin, which locks up coconut and citrus-peel flavors. But all that is academic, because the only reason anyone grows Kingston Black is for the mysterious flavors that emerge when its juice is fermented. It offers a perfect balance of sweet, tart, bitter, and savory saddle sweat, like prosciutto. **Texture** A bit on the mushy side, but who cares? **Season** Pick in November. **Use** Hard cider. **Region** Revered by Somerset cider makers and upstart American ones.

Some apples give off a vibe. "Kingston Black" sounds like the name of a hard-boiled private eye, the kind who calls women dames and slugs men before calling them anything. It is considered the most gifted cider apple in the world, yet the most difficult. The general rule with ciders is that you need to blend different varieties together to achieve harmony, but Kingston Black is a loner. It doesn't play well with others. It does, however, make one of the most distinctive single-variety ciders in the world, and for that it is highly sought. Robert Hogg, writing in 1851, considered this "beautiful little apple" to be "the most valuable cider apple" in England.

This truffle of the cider world has a nose that starts with roses, transitions to fruit, and ends with racy "happy sweat" aromas, or, in the shorthand of Farnum Hill Ciders (the only cidery with a significant supply of the fruit), PWLEO: People Who Like Each Other. As Nicole LeGrand Leibon, Farnum Hill's cider maker, puts it, "It's like a good date, from beginning to end." But note that one person's "happy sweat" is another's "rubbery/cheesy/barnyard"—typical terms used to describe the sweeter, funkier style of Kingston Black cider traditionally made in England. Whether you meet it in England or America, you'll remember Kingston Black's killer finish, which makes for an amazingly long good-bye.

Medaille d'Or

Origin Bois-Guillaume, Rouen, Normandy, France, 1800s. **Appearance** Lovely little oblong golden apples webbed in a fine russet patina. These look like something you'd pick up in an antiques store. **Flavor** A tapenade of soft tannins. You can eat it fresh and still have a good time, the tannins tickling the bitter taste buds on the middle of your tongue like olives do, without any of the back-of-the-tongue thrashing you get from hard tannins. The strong astringency will have the tip of your tongue flicking over your lips, feeling the fuzz. **Texture** Wonderfully tender without being mushy. **Season** Late fall. **Use** Hard cider **Region** Normandy.

On a list of thirteen classic cider apples in *The Story of the Apple*, Medaille d'Or has at least twice the tannin content (0.64 percent) of any of the others, save Chisel Jersey, a distant second at 0.40 percent. Medaille D'Or packs a wallop. Yet in my experience it isn't nearly the spitter that some other cider apples are; it just gives a lot of that fuzzy-lips feeling. It is also very high in sugar, a classic "bittersweet," which has always made it an important apple in Calvados and other Norman brandies. It was imported to England in 1884 and used for cider there, too. It makes intensely rich and winelike cider with a long finish. The apple was first discovered in Bois-Guillaume, a snooty little enclave atop a hill in Greater Rouen where William the Conqueror's mom built her summer place in 1040. Don't call this crotchety Gaul "Gold Medal," which sounds like some cloying, breeding-program spawn of Golden Delicious.

Redfield

Origin Geneva, New York, 1938. (Cross of Wolf River and Niedzwetzkyana.) **Appearance** A hearty, vivid red, Golden Delicious–size apple. Inside, the flesh is red beneath the skin and white in the middle strip, with a blood-red heart. **Flavor** Very tart and aromatic, like a sour cherry. **Texture** Crisp and firm, with chewy skin. **Season** Late fall. **Use** Makes a crazy pie. Achieves greatness in hard cider. **Region** Grown by a handful of aficionados in the Northeast.

You know something's up with a Redfield apple tree in spring, when it unveils startling pink petals etched with white squiggles. That color is made by anthocyanins, the same pigments that make raspberries red, and Redfield trees are veritable fountains of anthocyanins, which turn the blossoms fuchsia, the leaves coppery, the apple skins rich red, and the flesh vivid pink. They also impart a cherry-berry flavor. People say Redfield is too tart and astringent for fresh eating, but don't tell that to my son, who likes the puckery ones.

Redfield is one of the least known but most enchanting graduates of that Hogwarts for apples that is the New York State Agricultural Experiment Station. Its particular wizardry involves creating gorgeous, dry, and aromatic blush ciders. West County Cider, of Shelburne, Massachusetts, has carried the torch for years, but now others are catching on. Redfield was developed by crossing the large, mild Wolf River with Niedzwetzkyana, the Redvein Crab of the Caucasus Mountains between the Black and Caspian Seas. (Niedzwetzkyana, which has freaky burgundy flesh, is the progenitor of all our red-fleshed apples.)

Although Redfield is not an old variety, it makes me think of the legend of Micah Rood. In 1888, the *New York Times* published a story about Micah Rood apples, which arrived in Connecticut markets each fall from the western town of Franklin. The apples had "cherry-red skin" and a "snowy interior," but their defining characteristic was "a large red globule near the heart of the fruit resembling a drop of blood." The orchardists of Franklin, nearly all of whom possessed some Micah Rood trees, explained that the apple was named for a Franklin farmer of the early 1700s. Micah Rood was a strange old coot, described by his peers as avaricious and indolent. One day a traveling peddler who had headed toward Rood's farm, looking for a spot to rest his head for the night, was found murdered beneath Rood's apple tree the following morning, his money gone. Although suspicion settled heavily on Rood, there was no evidence—until the following autumn, when every apple on the tree was found to have a "bloody heart." This made Micah Rood more crabby than ever, and his life went steeply downhill.

You'd think that the pall of sin hanging thick on the tree would have made the good people of Franklin shun the bloody-heart apples, but they did just the opposite, perhaps in part because the apples were unusually delicious. They eagerly grafted the original tree, and soon every farmer in Franklin was growing the curiosities—and promoting the legend. The apple itself seems to have disappeared sometime after 1888, but now I have grafted the tale onto the stem of the Redfield, where with luck it will take.

Wickson

Origin Humboldt County, California, 1944. **Appearance** Like a Rainier cherry on steroids, this Lady Apple–size shrimp features swirled red and yellow quadrants and a long stem. **Flavor** Wicked tart, wonderfully perfumed, wildly sweet. Wickson's off-the-charts acidity disguises its concentrated sugars. This all ferments beautifully into a bone-dry, water-white, high-alcohol cider with a nose of guava and lychee and an astringent crab apple finish. **Texture** Slightly fibrous, like Hewes Crab, and incredibly juicy, making it excellent for pressing into a nice, clear cider. Wickson's very thin skin means the apple doesn't keep. **Season** October. **Use** Wickson's hobbit proportions can make it a bit frustrating to eat, though its one-two punch of sugar and acid makes it a delight for thrill-seekers, as well as the best cider apple introduced in a century. The ideal candidate for Slow-Roasted Baby Apples (page 260). **Region** Never embraced by the industry, but now a hit with small-scale cider makers.

This little knockout punches above its weight. With a ridiculous sugar content of 25 percent, and acid to match, and all sorts of crab apple weirdness going on, Wickson can test your limits. It stands as the masterwork of Albert Etter, the crackpot California pomologist who ran his Humboldt County gardens like a mad scientist's lab in the early twentieth century. (See "Pink Pearl," page 248, for the full story.) Were it not the size of an underachieving kiwi, Wickson might have gone on to glory soon after Etter released it in 1944. But its size suited it only to cider making, and nobody was looking for great cider apples in 1944. It lingered on the curio rack until a handful of cider makers discovered its virtues in the 1990s. Foremost among them was Steve Wood, who uses Wickson to give Farnum Hill ciders their distinctive long, tart finish. In his words, "You should want another glass." Straight up, Wickson cider will score your gums, but a little Wickson in a blend boosts alcohol and acidity, helping to preserve and brighten the cider. Etter named this fruit for E. J. Wickson, the father of California pomology, and one of the few people who didn't think Etter was off his rocker.

Yates

Alias Red Warrior **Origin** Fayette County, Georgia, 1840s. **Appearance** Nice, shiny red skin with pale dots that seem to be cascading toward the flower end. Some russet around the stem. Yates tends to stay quite small and flattened unless well thinned. **Flavor** Sweet and slightly tart, a southern take on that spicy, cidery Mac twang. Good eating. **Texture** Snappy and so juicy it sometimes squirts when you bite into it. The skin is fairly tough. **Season** October in the South. Keeps forever. **Use** Cider, and storage into spring. **Region** Deep South.

One of the most important southern apples, Yates can take full southern heat and humidity (as you'd expect from its Georgia origins) and still keeps incredibly well, which made it extremely useful prerefrigeration. Its small size lent it to cider in particular—Southerners would typically store the apples away until other cider varieties had been exhausted, then press the Yates in winter or even spring.

ODDBALLS

Most dog breeds made it into humanity's good graces by doing something useful. They could herd sheep, or track foxes, or pull a cart, or play nice with children. Then there are the Shar Peis of the world, who made it into the canon on weirdness alone. The apple is no different. If enough people want you, for any reason, you will survive; this last group of apples includes those whose desirability has little to do with flavor or utility. Instead, they have come up with a variety of tricks to make themselves irresistible to apple collectors: by being small and cute, for example, or having red flesh, or by being the ugliest apple on the planet, or simply by being linked to some momentous bit of history. Throughout this book I've been making the case for appreciation of the apple's extraordinary genetic creativity; consider this final section my closing argument.

Api Étoile

Alias Star Lady Apple **Origin** Switzerland, 1600s. **Appearance** You will never forget this tiny apple, which is so ribbed it resembles a five-pointed star. Like its close relative the Lady, its light green skin and pink cheek are shellacked in the shiniest wax you've ever seen. In a store, you might think these were fake. **Flavor** A sprightly, sweet-tart zing is wasted, because these really aren't very easy to eat. **Texture** Dense, tough, juicy, very hard to gnaw on. You feel like a rat, forced to use your incisors. The amazingly thick, glossy skin retards moisture loss, which is why these apples keep forever. Strangely dense and remarkably heavy for the size. **Season** Late fall, just in time for the holidays. **Use** Ornamental. They resemble wax fruit, and last almost as long. They would actually make a great fresh cider, but that seems like a waste of anything this pretty. Save them for your wreath. **Region** Rare. Found only in a few collectors' orchards.

Why isn't the "Star Lady" a ubiquitous presence in every high-end supermarket come the holiday season? What store would not sell out of these little gems? Lady apples are cute and all, but these Lady offshoots could take the edge off the sourest Scrooge. Resistance is futile.

Flower of Kent

Alias Isaac Newton's Tree **Origin** Kent, England, 1629. **Appearance** Huge, blocky, acutely ribbed, elongated, and quite heavy. A commanding presence, and a handsome one with its carmine-burgundy stripes. **Flavor** Sour. **Texture** Soft and mealy. **Season** September. **Use** A traditional English pie apple, meaning it dissolves into a fine puree, meaning it makes a lovely sauce. The size and the softness make it problematic for out-of-hand eating. **Region** Very rare in England, powerfully rare in the United States. Specimens can be found at Woolsthorpe Manor, the University of York, Kew Gardens, and the USDA's apple collection in Geneva, New York.

In the summer of 1666, twenty-two-year-old Isaac Newton was sitting in the garden of Woolsthorpe Manor, his family estate in Lincolnshire, killing time while the plague swept through London, when he saw an apple fall to the ground. (Contrary to popular thought, the apple did not bean him.) The gears started whirring, and soon he hit upon the notion that gravity was a force of attraction between two objects, and that the more massive object would attract the lighter one, and that the same force acting upon the apple might also be acting upon the moon. That tree was a rare Flower of Kent, and, unbelievably, it still exists at Woolsthorpe Manor. Although an 1816 storm felled the tree, one branch rooted itself, and today a full-size tree can be seen growing out of the old carcass. All Flower of Kents in existence stem from Newton's tree. The apple itself is tart and coarse, which is its great contribution to science, since a tastier fruit no doubt would have been plucked from the tree long before it had the chance to fall so weightily.

Hidden Rose

Alias Airlie Red Flesh **Origin** Airlie, Oregon, 1960s. **Appearance** A small, green, conical, prim apple, sometimes with one red cheek. Its waxy, translucent skin is speckled with white dots rimmed in green, like frog eggs in a pond. Floating beneath the surface is the faintest wash of rose, a hint of what's inside (in case the name didn't clue you in). Sliced open, the fruit is shockingly pink, only the inner sanctum staying white. **Flavor** Like the world's crunchiest strawberry, tart and juicy, delightfully light and refreshing. **Texture** Crisp and juicy when young, but it turns mushy if not kept cold. **Season** Ripe in September to October. **Use** Display in slices on a smorgasbord, since it stays pink instead of turning brown. A kid fave. Makes zany sauce. Passable in pie. **Region** Concentrated in the Pacific Northwest, but turning up everywhere.

Some experts call Hidden Rose the best of all red-fleshed varieties, and predict greatness for it in the market. It is a chance seedling discovered in a field in Airlie, Oregon, in the 1960s. For a while, it was suspected of being a Pink Pearl, but in the 1980s it was shown to the pomologist Bill Schultz, who pronounced it a unique variety and named it Airlie Red Flesh. (Still, it seems closely related to Pink Pearl and the other red-fleshed apples bred by Albert Etter in California in the early 1900s, one of which may well have been the pollen parent.) It was later trademarked as Hidden Rose by the farm that owned the tree. Sweeter and with a crisper texture than Pink Pearl, it won't stay hidden for long.

Kandil Sinap

Origin Sinop Peninsula, Turkey, early 1800s. **Appearance** Unmistakable. Like a golden vase splashed with carmine on the sunny side. Very smooth and waxy, with a barely noticeable, orderly grid of lenticels. **Flavor** Sweet and perfumed, this is the Turkish delight of apples. **Texture** Crisp and juicy, fine grained and tender, with a thick, chewy skin. **Season** October. Lasts into early winter. **Use** Best eaten fresh. **Region** Still big in Turkey. Popular as an oddity among fruit collectors worldwide, but it does best in a warm climate.

Kandil Sinap wobbles like a Weeble and often falls down. In the mini category of crazy elongated apples, such as the Glockenapfel and Sheepnose, Kandil Sinap is the most extreme. The apple's exotic beauty actually distracts from its superb flavor; if it was dumpy and dull, everybody would rave about the taste. This native of Sinop, a province on Turkey's Black Sea coast (the apple name means "Sinop candle," a reference to the shape), seems like the kind of apple that would have been reserved for sultans back in the days of the spice trade. Any commoner who set eyes on one would have been instantly executed. Is there any doubt that the king of Persia idly snacked on imported Kandil Sinaps as he listened to Scheherazade spin her tales?

Kazakh Wild Apples

Origin Tian Shan Mountains, 10 million years ago. **Appearance** Small, but otherwise covering the entire spectrum of apple colors, patterns, and shapes. **Flavor** Sour, bitter, sweet, and mysterious. **Texture** Sometimes crunchy, sometimes soft, rarely crisp like a modern apple. **Season** August to September. **Use** Best for sauce and cider. **Region** Tian Shan Mountains of eastern Kazakhstan and western China. Also Geneva, New York.

The birthplace of all apples is the forested flanks of the Tian Shan, the Heavenly Mountains of Kazakhstan. Although decimated by deforestation, those wild apple forests survive. In places, two-thirds of the trees are apples, with a smattering of apricots mixed in. Imagine wandering these forests of the Heavenly Mountains in fall, the canopy and ground filled with colorful jewels; it's about as close to the Garden of Eden as any of us will ever get.

Since every apple tree that grows from seed is a unique variety, these forests are home to literally millions of new "varieties," and it shows. Here, all the possibilities of appleness mingle and rebound off one another like genetic billiard balls. There are tiny green and medium green apples, yellow minis and sizable red ones, pint-size Gravenstein lookalikes and bitter green spitters, a baby Winter Banana, some yellow-fleshed, a few where the red bleeds from the skin into the flesh, and some that could do a pretty good impression of a Knobbed Russet. The apples tend to be small, because no one is pruning the trees, exposing them to full sunlight, or thinning their fruit, but the genetic genius of the apple is on full display.

In 1989, Phil Forsline, curator for the USDA's Plant Genetic Resources Unit in Geneva, New York (home to the largest apple collection in the world), led the first of three expeditions to the apple forests of Kazakhstan to bring some of that diversity back to America, and potentially reintroduce some genes with disease resistance, the ability to withstand drought or heat, or other useful tricks that had been lost to the apples we domesticated. They collected tens of thousands of seeds, along with scion wood, and started a wild Kazakh orchard in Geneva. Today that orchard is an unruly forest of twenty-foot trees festooned with multicolored fruit (the source for the photo here). Wandering its aisles in autumn, inhaling the bewitching aromas, is like stepping back in time, back into the Heavenly Mountains.

Knobbed Russet

Alias Knobby Russet **Origin** Sussex, England, 1819. **Appearance** "It's alive! … It's alive! … It's alive!" The russet skin is always covered with raised welts that resemble scars. The greenish yellow background is veined pink, like a drunkard's nose. **Flavor** Surprisingly good. Sweet and nutty, with a fino sherry tang. **Texture** Strangely brittle, almost like a raw potato. Very firm, but you wouldn't call it crisp. There is almost no grain to the flesh, which makes it smooth, almost slippery. The welts are woody and inedible. **Season** October. Like most russets, it's a good keeper. **Use** Terrify your children. It's good fresh and excellent (peeled) in crisps. **Region** Although not common anywhere, its primary lair is eccentric orchards throughout the United States and the United Kingdom.

There is nothing wrong with this apple. This is how it looks when everything goes *right*. It seems as if it stepped straight out of an H. P. Lovecraft tale, and it tends to elicit creative descriptions from people. "Looks like a rotting brain!" is not uncommon. The welts and warty knobs discourage a direct mouth attack, but once you peel off the skin (surprisingly easy), you'll discover a sweet, funky, tasty beast, both perfumed and earthy at the same time. But let's face it, this apple has not been cultivated for two hundred years because of its flavor; it is here to freak out your friends.

Lady

Aliases Api, Pomme d'Api, Christmas Apple, Longbois **Origin** Brittany, France, 1500s. **Appearance** The world's cutest apple; two bites and you're done. You can easily hold three or four of these doughnut-shaped jewels in your hand. The Lady in fall is bright green with one perfectly embarrassed cheek. By Christmas, she has turned pale yellow on one half and scarlet on the other. **Flavor** One of the most delicious apples, Lady is bright, nutty, and fruity, with a blueberry-like finish—astringent and rich. By Christmas, the acid fades and she becomes sweet as pear drops. **Texture** The flesh is fine and white, but the nature of the Lady is that the diminutive size ensures that you are eating a large ratio of skin to flesh. Fortunately, the skin is nice and tender. The core is noticeably small in proportion. **Season** Christmas. **Use** Holiday decoration. Also a great fresh snack. **Region** Throughout Europe and the United States.

An ancient apple, with a special connection to Christmas that has been fairly unchanged for hundreds of years. The French (and, until fairly recently, the British) refer to this apple as Api, and there has been some conjecture by authorities that this is the same fruit called *appiana* by Pliny the Elder in the first century A.D. This is why you'll see many claims that this same apple was eaten in Ancient Rome. The idea seems pretty weak to me; more likely, *appiana* has something to do with Appius Claudius Caecus, builder of the Appian Way, the first highway to southern Italy. The other story, more believable, is that the Api was first discovered as a seedling in the Forest of Api in Brittany. No date is given for this discovery, but it would have been no later than the 1500s, for by the early 1600s the Api starts to turn up as an important apple for French royalty. Jean-Baptiste de la Quintinye, chief gardener to Louis XIV, adored the Api and put it to great use in Versailles feasts. He recommended that it be eaten "greedily, and at a chop; that is to say without ceremony and with its coat all on, for none have so fine and delicate a skin as this." Quintinye used it for Christmas table decorations and in wassail bowls—proving that this little jewel's association with Christmas is not a recent invention.

One seventeenth-century British writer described the Lady as the apple "which the Madams of France carry in their pockets, by reason they yield no unpleasant scent." (More likely a breath freshener than a deodorant, methinks.) Another noted the apple's habit of being very pale on one side and very red on the other, and theorized that it "may serve the ladies at their toilets as a pattern to paint by." This scarlet-cheeked tendency, I suspect, is why Americans started calling the apple Lady.

By the 1840s great shipments of Ladies were being sent from New York to London for the Christmas season, where it sold at a higher price than any other apple. Hogg described how, "in the winter months, they may be seen encircled with various coloured tissue papers, adorning the windows of the fruiterers in Covent Garden Market." The same was true in Manhattan, and is true still.

Few sights are prettier than a Lady tree full of apples. The fruit hangs tight to the tree's long, upright branches (hence the nickname Longbois) like natural garlands, thirty or forty to a branch. This is also my dog's favorite apple. He's small, and one fits perfectly in his mouth. When we use Lady apples in displays around Christmas, he gazes forlornly at them and whimpers until we give him one.

Pink Pearl

Origin Humboldt County, California, 1944. **Appearance** Waxy, pale, pearlescent skin, stippled with brown dots ringed in green. Acutely ribbed, with a few sprays of russet emerging from the stem. Everywhere, the pink interior peeks through. A bite reveals what looks like cherry-vanilla ice cream inside. **Flavor** An assertive berrylike tartness, like the tiniest raspberries that never sugar up. Some say grapefruit, though I haven't detected it. **Texture** A bit mealy, but tender and juicy. The skin is like gossamer. **Season** August in California, September elsewhere. **Use** Eat fresh for the novelty. Makes a lovely pink applesauce. The fresh cider would probably be gorgeous and lip-smacking good, but I've never seen any. **Region** Mostly California and the Pacific Northwest.

In 1894, a twenty-two-year-old Californian named Albert Etter claimed a parcel of Humboldt County wilderness through the Homestead Act and transformed it into Ettersburg, a sort of Island of Dr. Moreau of plant breeding. In contrast to the experts of the day, who favored slow, conservative tweaking of the gene lines in pursuit of modest improvements, Etter liked slamming two exotic individuals together to see what they might engender. Ignoring the derision of the scientific establishment, Etter made thousands and thousands of crosses in his experimental gardens at Ettersburg. He created a lot of freaks, but also many strange and intriguing beings, including some of the world's best strawberries and some striking apples. Etter had a particular fascination with red-fleshed apples. Using a pink-fleshed variety called Surprise, which was a descendant of Niedzwetzkyana, the original red crab apple of the Caucasus, he made countless crosses and eventually chose thirty selections for potential release. Of those, only Pink Pearl caught on with growers. (Ettersburg fell into ruin after Etter's death, though many of its creations survive in the collection of Greenmantle Nursery.) The vibrant pinkness of this apple's flesh glows through the translucent green (later yellow) skin in stripes, making Pink Pearl look more like a half-ripe tomato than an apple. The crazy stripes of red and white make it great fun to eat. The Pink Pearl never sugars up much, but some take a liking to its sour berry zip.

Red Delicious

Aliases Delicious, Hawkeye **Origin** Peru, Iowa, 1881. **Appearance** Unmistakable. The image of the conical, ribbed, midnight-red Red Delicious, on its five-point stand, is burned into every American's brain. It really is a handsome apple, with white lenticels emerging against that midnight-red background. **Flavor** Remarkably bland, even when grown well. A hint of watery sweetness and that's it. No acidity to speak of. The skin has a bitter arugula flavor. **Texture** A good, tree-ripened Red Delicious is actually pretty crisp and juicy, though far too many are dry and cottony inside. Both good and bad examples have that horribly leathery skin that likes to slide between your teeth and lacerate your gums. **Season** Picked in September and October, then held in controlled atmosphere storage and sold year-round. **Use** Makes a great logo. **Region** The national apple, sold in every supermarket. Most are grown in Washington State.

Red Delicious, the most popular apple in America, lumped in with the Oddballs? Well, it certainly can't be recommended for culinary purposes or for eating out of hand, and it isn't that great a keeper. What defines it is appearance: a freakish, deep-burgundy molar. It is more icon than fruit.

Yet it wasn't always so. The apple that would become the Red Delicious began life as a sprout beneath a Yellow Bellflower on the Iowa farm of a Quaker named Jesse Hiatt. Hiatt took one look at the tree that would one day rule them all, and promptly cut it down. It came back up, so he cut it down again. Yet the apple was not to be denied its dharma; it sprouted a third time, which the Quaker took as some sort of sign. He welcomed it as part of his orchard after that, even though it wasn't in any of the rows. A few years later, it produced its first sweet, perfumed, and crisp fruit. Hiatt tasted greatness, and promptly named it ... the Hawkeye. (He was a good Iowan, after all.) The Hawkeye was amazingly sweet, and by 1893 Hiatt was so convinced of its brilliance that he sent a few samples to Missouri, where Stark Bro's Nursery was holding a contest to find the next great apple to replace Ben Davis, then the dominant apple of the Midwest, which (ironically, as things turned out) was a beautiful red apple that tasted like cardboard. The Hawkeye won the tasting, but Hiatt's contact info had been misplaced; no one knew where the winning apple had come from. Fortunately, Hiatt entered the same fruit show the following year, and Clarence Stark recognized the apple, bought the rights from Hiatt, and dubbed it Delicious, a name he'd been saving for just the right apple.

Hiatt died a few years later, but Stark Bro's sold millions of Delicious trees. The Delicious was a fantastic grower and producer, and the fruit kept well and had an inoffensive, pleasantly aromatic taste. Most of all, it was very sweet. What it wasn't was solid red; instead, it had a light pink blush, reddish stripes, and a less pronounced strawberry shape, making it a pretty generic apple. That heirloom, the Hawkeye (on the right in the portrait), can still be found in certain orchards.

In 1914, when Stark Bro's released its Golden Delicious, the name was changed yet again to Red Delicious. As the Washington State apple industry began to swamp the rest of the country during the twentieth century, Red Delicious became the most successful apple in the history of the world, largely because of its shipping qualities. Bunyard called it "the best of the red apples which cross the Atlantic," which was faint praise, since he had contempt for the other red apples crossing the Atlantic.

During the 1950s, Red Delicious took over America. It became America, merging with other iconic symbols to form the image of the

post–World War II, clean-living goliath. It also became redder. Red Delicious has a tendency to produce sports—single branches with genetic mutations. These sports can be grafted just like any other variety. Over time, redder and redder Red Delicious sports were selected by growers—and well they should have been. Traditionally, growers were paid based on the redness of the skin of their apples. Flavor was not evaluated. Red Delicious earned a premium over other apples, and the reddest Red Delicious earned the highest premium. That is how we wound up with an apple that looks like it was painted deep burgundy and shellacked.

It has become popular to blame a conspiracy of growers and wholesalers for this. As national supermarket chains came to dominate the grocery business, they required more and more standardization from their suppliers. They wanted identical apples in every store, and they wanted to buy those apples from as few people as possible. The efficiency of scale, the industrial model, took over. The nation's apple section turned into one gigantic bin of Red Delicious because that is what made the most money for the major players.

But really, the enemy was us. That whole system could work only if consumers bought the apples—and we did. One generation removed from the farm, we had lost our apple smarts. We simply chose the reddest apples we could find, associating redness with ripeness. The supermarkets, wholesalers, and growers learned their lesson well, and kept providing us with what we'd clearly shown we preferred. Red Delicious picked while deeply red but horribly unripe (which made them keep even longer) outsold other apples.

But, like a yin-yang symbol, within the heart of that increasing domination grew the seed of its own destruction, because as Red Delicious got redder and redder—the industry's term of art is "midnight red"—it got more and more tasteless. Partly what made the sports redder was thicker skin—a deeper canvas to paint on—and as the redder sports were selected, the genes that produced sweetness got left behind. Consumers may be gullible, but they do learn eventually. At some point, the Red Delicious became uncool.

Although it feels like it has always dominated the apple scene, Red Delicious's reign was surprisingly short. It was not until the mid-1960s that it produced 50 percent of Washington State's apple crop, and the mid-1980s that it reached 75 percent. By then, savvy consumers were switching to Granny Smith, or whatever they could find locally, and the price paid for Red Delicious plunged. It became the cheap, mass-market apple, and Washington State growers stopped planting it, chasing the consumers to Grannies and Galas and Fujis and, now, Honeycrisps.

Today, Red Delicious is a zombie apple. Although still ubiquitous in the marketplace, the life left it long ago. It has a zombie taste, and a zombie future. It may still be America's number-one apple, with 40 percent of the market, but no one has planted Red Delicious in years. In two decades, it will be as forgotten as Winesap is today—and far less mourned.

Sheepnose

Aliases Black Gilliflower, Gilliflower **Origin** Connecticut, late 1700s. **Appearance** Uniquely appealing, the Sheepnose is eccentric, elongated, and pointy nosed, like its namesake. The skin turns a handsome matte burgundy, sometimes nearly black, over a green background. **Flavor** Jolly Rancher meets watermelon rind, sweet but lacking in acidity. Apparently some people have a recessive gene that detects the strong flavor of clove in these apples. Unless my "Jolly Rancher" is your "clove," I don't have the gene. **Texture** The very definition of mealy. Virtually juiceless. Like sinking your teeth into an old baked potato. **Season** October. Not a keeper. **Use** Sauce? Cider? Dinner party conversation piece? Some people like them straight off the tree. Some people are weird. **Region** Quirky New England orchards.

In your average tasting of heirloom apples, this ranks dead last. It actually has decent flavor, but that's beside the point, because it's almost impossible to get beyond the texture. The best that Beach could say about it in 1905's *Apples of New York* was "the flesh at its best is but moderately juicy and soon becomes dry, but it has a peculiar aroma which is pleasing to many." That aroma is indeed both peculiar and pleasing, yet it isn't enough to have saved this apple from oblivion for two hundred years. It was already "fast becoming obsolete" in 1905, yet somehow it has hung around, and it turns up in many heirloom orchards. Maybe the elusive clove scent drives a certain segment of the market wild. Gilliflower is an old name for the carnation, which was once used in Spain to give drinks a clove flavor. Zeke Goodband believes Sheepnose is the secret ingredient that makes his fresh cider at Scott Farm so spicy and irresistible.

Winter Banana

Alias Banana Apple **Origin** Cass County, Indiana, 1876. **Appearance** A stunner, with waxy, banana-yellow skin, a pink blush on whatever cheek faced the sun, and dark freckles. This large apple feels even heavier in the hand, due to its density. Sometimes it has a prominent suture mark. **Flavor** There's banana here, yes, but it's those little exotic bananas that can taste powerfully green and tannic when not quite ripe. A strange, enjoyable, musky flavor. Adzuki bean comes to mind. With time, the astringency fades, leaving a gentle lemon-curd richness. **Texture** Starts life very hard and woody, then softens in storage. **Season** Pick in October, store until November to December. Despite the name, this thin-skinned apple doesn't keep exceptionally well. **Use** Good for eating fresh in early winter. Its shape does not soften in pies, but its flavor disappears. **Region** Nationwide, especially low-chill zones like California and the Deep South.

Winter Banana's popularity rests on four pillars. It is lovely to behold. It has a winningly peculiar name. It needs no winter chill to initiate flowering. And it is one of the best pollinators of other apple varieties, and is often planted amid orchards for that very purpose. Note that "taste" is not on this list. Winter Banana has an alluring, if mild, perfume, but it wouldn't stand out on taste alone. Yet the whole package—it will feed you well in December, it will help your other trees to raise their game, and it will class up any fruit basket—makes it a perennial favorite.

RECIPES

The goal of these recipes is to lower the barriers to apple usage, and to try to capture some of that amazing diversity that the apple has displayed throughout this book. Apples make superb appetizers, salads, entrees, side dishes, drinks, and, of course, desserts. They go well with a number of different cuisines. They are awfully easy to work with.

One thing you'll quickly notice is that none of these recipes call for peeling the apples. I suspect that our instinct to automatically peel apples dates from the dark days when the market was dominated by thick-hided brutes like Winesap, Red Delicious, Granny Smith, and McIntosh. The lion's share of the modern apples on the market today, and quite a few of the heirlooms, have tender skins that deepen the pleasure and complexity of most dishes. Most of us who've stopped peeling our apples can't imagine going back.

Besides, when you don't have to deal with peels, you can have an apple cored and sliced or diced in seconds, and that makes it all the easier to cook with apples frequently. So, if you really hate apple peels, then by all means peel your apples, but you may find, as I have, that all that peeling was making your food a little less interesting and your life a little harder.

You may also find that you prefer most of these recipes with a mix of apples. That's certainly the secret to a good pie: a mix of tart apples and sweet ones, some crisp apples for bite and some tender ones to glue the whole pie together, and a few exotic specimens for intrigue. It's essential for pie, but it's also true in most other cases.

I've included some suggestions of apples that would work well in each dish, but that is all they are. No dish requires any particular variety of apple; just make sure to use a firm apple when you need the fruit to keep its shape, a tender apple when you want it to dissolve into sauce, a tart apple when you need to liven things up, and a sweet apple when the dish needs no tang. Play around with different combinations; it will all be good.

SLOW-ROASTED BABY APPLES

Works with any small apples, but best with tart and tannic apples, where the sugar offsets the edgy qualities. If you use **Ladies**, which are very sweet, you need little sugar. **Wicksons**, with their intense contrast of sweet and sour and their uncommonly long stems, would be dreamy.

40 small apples, stems intact

1/4 cup canola oil

1 tablespoon cinnamon

2 tablespoons sugar

If you should ever find yourself in the wild apple forests of Kazakhstan with time to kill, you'll be glad you have this recipe with you. It's the solution for what to do with all the tiny apples of the world, and it works well with wild apples, roadside specimens, large crab apples like Dolgo, and even Lady or other small commercial apples. You can also roast dessert apples during the summer, when they are small and sour. The beauty of it is that there's so little prep. The apples roast to a beautiful, bronze crinkliness, the sugars caramelize, and the stems serve as handles: you lower each one into your mouth and pull off the flesh with your lips, leaving the core attached to the stem. You can roll them in sugar, as below, and delight groups of small children (or anyone's inner child), or skip the sugar and bake them around a roast. Or serve small bowls of them to guests to accompany Pomona Sidecars (2 parts apple brandy, 1 part Cointreau, 1 part lemon juice, shaken well with ice and strained into a cocktail glass that has had the rim rubbed in lemon juice and dipped in sugar).

1. Preheat the oven to 300 degrees.
2. Place the apples in a large bowl and toss with the canola oil.
3. Add the cinnamon and sugar and toss the apples to mix well.
4. Place the apples on a baking sheet and roast until soft, at least 1 hour. You want them soft and wrinkled but not collapsed. Serve in low bowls or wide cocktail glasses.

Makes 6 to 10 servings

APPLE PAKORAS WITH YOGURT-MINT CHUTNEY

This might be a good spot for the modern, crisp, foam-textured apples, like **Honeycrisp**, **SweeTango**, and **Pixie Crunch**. **Wagener** would give you tender, spicy pakoras, while **Northern Spy** or **Winesap** would give you tangy, snappy ones. **Keepsake**: Yum!

For the Chutney

2–3 cups loosely packed mint leaves

1 cup plain whole-milk yogurt (preferably Greek)

1 garlic clove

Pinch of cayenne (optional)

1-inch piece of ginger, peeled

1 teaspoon salt

1 teaspoon coriander

¼ teaspoon cardamom (optional)

Juice of ½ lime (optional)

For the Pakoras

2 cups chickpea flour

1 teaspoon baking powder

1 teaspoon salt

1 tablespoon garam masala

1 teaspoon cinnamon

1 cup water

2 large apples, cored and diced

Canola oil for frying

Mint leaves and/or flower petals for garnish

A fragrant apple fritter that brings the apple back to its Asian roots. Garam masala is a spice mixture of the "four Cs"—cardamom, cinnamon, clove, cumin—plus sometimes a fifth, coriander, and sometimes others like mace, nutmeg, or ginger. Like Vishnu, it appears in myriad different guises, so don't sweat the details; use whatever mix of these you have on hand.

1. Make the chutney: Combine all the chutney ingredients in the bowl of a food processor and blend. Leaving a little texture to the mint is nice. Refrigerate until ready to use. (The flavors actually meld better if you let it sit for a few hours.)
2. Make the pakoras: Combine the chickpea flour, baking powder, salt, garam masala, and cinnamon in a large bowl.
3. Add the water and mix well. The batter should have the consistency of pancakes or heavy cream. If it seems too thick, add more water.
4. Stir in the apples.
5. Heat the oil in a heavy skillet over medium heat. The oil needs to be at least 2 inches deep so the pakoras can float, and 3 inches is better.
6. When a pinch of flour sizzles as soon as it touches the oil, you are ready to fry. Spoon the batter into the oil one large spoonful at a time. Spoon in as many as can fit without crowding one another. They will float to the surface and sizzle.
7. Cook until they get a nice, dark brown color, 4 to 5 minutes, turning once around the 3-minute mark. Don't be fooled; light golden is not quite done.
8. Lift the pakoras out of the oil, drain on paper towels, garnish with mint leaves and flower petals (if you want) and serve at once with the chutney.

Makes 4 to 6 appetizer servings

CHICKEN LIVER PÂTÉ WITH APPLES AND CALVADOS

Firmness is not an issue here, so sweet and flavorful are the primary requirements. All the modern apples—**Fuji**, **Gala**, **Ambrosia**, **Jonagold**, **Pink Lady**, **Pinova**—would be great, as would **Baldwin**. The perfume of **Maiden's Blush** or **Rome** would do wonders here.

2 tablespoons bacon fat

3 large shallots, chopped

1 large apple, cored and chopped

1 pound chicken livers, trimmed of any weirdness

Salt and pepper

¼ cup Calvados or apple brandy

8 tablespoons (1 stick) **cold butter, cut into 6–8 pieces**

I make this every December, freeze it, and hand out little ramekins of it through the holidays. It's more appley and less livery than typical pâté, and has thus converted many a pâté skeptic. If you don't have cans of bacon fat around, just substitute butter. Serve with toasted baguette slices, tart sliced apples, and a glass of sparkling cider. Also consider a deconstructed version of this as a pasta dish: keep the sautéed apples and shallots separate, toss the creamy liver sauce in linguine, and top with the apples and shallots (and bacon, perhaps). For a truly silky texture, you'll need to peel the apple.

1. Heat the bacon fat in a large skillet over medium heat. Add the shallots and sauté, stirring occasionally, until soft, about 3 minutes.
2. Add the apple and cook an additional 2 minutes.
3. Add the chicken livers and cook, stirring occasionally, an additional 2 minutes, until they are brown on the outside but still pink inside. Season with a little salt and a lot of black pepper.
4. Add the Calvados and cook, stirring, until the sauce is bubbling and has cooked down slightly, about 2 minutes.
5. Scrape the contents of the skillet into the bowl of a food processor and blend until smooth. With the motor still running, drop the butter down the feed tube one piece at a time so that it emulsifies. As soon as the last piece of butter has been incorporated, turn off the motor and pour the pâté into a series of ramekins or small bowls. Cover each container with plastic wrap, pressing the wrap right into the surface of the pâté, and refrigerate until cold, or freeze until needed.

Makes 6 to 8 servings

GRILLED APPLES WITH SMOKED TROUT, FENNEL, AND LEMON ZEST

I like to use one red apple and one green apple, putting one slice of each on each serving plate. Try apples that are both sweet and tart, and that are juicy enough to not dry out on the grill. **Honeycrisp**, **Jonagold**, **Gala**, **Rhode Island Greening**, **Karmijn de Sonnaville**, **Esopus Spitzenberg**, and **Newtown Pippin** would all be great.

4 ounces smoked trout

2 large apples, one red, one green

1 tablespoon canola oil

4 ounces sour cream or crème fraîche

Zest of ½ lemon for garnish

Fennel fronds, for garnish

Balsamic vinegar (optional)

A killer app: quick to make, lovely to behold, and perfect with a glass of tart sparkling cider. I like the gentle creaminess of crème fraîche here, but there's something to be said for the extra tang of sour cream, and sour cream definitely holds up better atop the hot apple rings. A slash of balsamic across the rings is both eye-catching and lip-smacking, but it may be gilding the lily. You decide.

1. Divide the trout into 8 equal pieces.
2. Preheat a grill.
3. Core the apples and slice them into 1/4-inch-thick rings. You should get at least 4 nice slices out of each apple, plus some ends that can be used or ignored.
4. Brush the apples with the oil.
5. Lower the grill heat. Grill the apples 3 minutes on one side, then turn and grill 3 minutes on the other side.
6. Remove the apple slices to individual serving plates (two per plate). Add a dollop of sour cream (off-center, please), lean a plank of smoked trout against it, and garnish with lemon zest and fennel fronds. If you are using the balsamic vinegar, simply drizzle a thin line across each ring. Serve immediately.

Makes 4 first-course servings

LOBSTER WALDORF SALAD

Now is the time to break out your most complex and tropical apples. **Pink Lady**, with its hints of mango and its resistance to browning, is perfect. **Cox's Orange Pippin**, **Fuji**, **Orleans Reinette**, **Pinova**, and **Zabergau Reinette** would all be excellent.

1 large apple, cored and diced

2 celery stalks, diced

½ cup peeled and diced celery root (optional)

¾ cup mayonnaise

1 ounce fresh tarragon leaves, minced

Salt and pepper

Lettuce leaves for garnish

Meat from 2 lobsters (kept in large pieces)

1 wedge lemon

Waldorf Salad (originally just apples, celery, and mayo; the walnuts came later) makes a great fresh, crunchy base for sweet chunks of lobster meat. I've tried this with both celery and celery root; I like the crunch of the celery, but the celery root has a more refined flavor. Consider a mix of the two. You can serve it over Bibb lettuce leaves as a salad, or stick the whole thing in a toasted hot dog bun for an elegant lobster roll. Either way, it makes a fine lunch. It would behoove you to make your own mayo here, but a little Hellmann's does the trick, too.

1. Mix the apple, celery, celery root (if using), mayo, and half the tarragon in a large bowl. Add salt and pepper to taste.
2. Line two medium salad bowls with the lettuce leaves. Divide the apple mixture between the bowls in two mounds.
3. Top each mound with half the lobster meat. Squeeze juice from the lemon wedge over the two salads. Sprinkle with the remaining tarragon and serve.

Makes 2 servings

CHORIZO IN CIDER, ASTURIAN STYLE

A crisp, sweet apple like **Gala** works, but a sweet-tart russet like **Zabergau Reinette** or **Belle de Boskoop** is even better. Any high-acid apple will do: **Bramley's Seedling**, **Esopus Spitzenberg**, **Ashmead's Kernel**, **GoldRush**, or **Granny Smith** in a pinch. **Ananas Reinette** might be best of all.

1 tablespoon olive oil

1 pound chorizo, cut into ½-inch slices

½ onion, halved and thinly sliced

1 cup dry hard cider

1 apple, cored and sliced into half-moons

Parsley for garnish (optional)

Bread for serving

Quick: Where can you find the strongest cider tradition in the world? Southwest England? Normandy? Actually, it's Asturias, the coastal province of northern Spain famed for its apple groves. Instead of sherry, *sidra* is the traditional drink in the tapas bars; walk into any bar in the region, and you'll be able to order this dish with your glass of cider. This is a bit of a jazzed-up version; more likely in Asturias, you'd get unadorned chorizo cooked in cider. I've made this recipe with dried chorizo, which is rock-hard when you buy it, and with fresh chorizo. Fresh is better, but you can use either one. You might even consider using a touch of dry in addition to the fresh, for intensity of flavor. Either way, serve it with crusty bread and a glass of cold, sparkling, funky cider; the dry, tart drink and the rich, spicy sausage play off each other brilliantly.

1. Heat the olive oil in a large skillet over medium heat.
2. Add the chorizo slices and sauté until brown, about 2 minutes. The oil will turn a lovely orange color from the paprika in the chorizo.
3. Turn the chorizo and sauté the opposite side another 2 minutes.
4. Add the onion and cider and continue to cook, stirring occasionally, for 6 minutes.
5. Add the apple slices, stir, and cook another 6 minutes, stirring occasionally, until the liquid has evaporated and the sauce is thick. Serve in a bowl, garnished with parsley if desired, and accompanied by thinly sliced bread.

Makes 4 to 6 first-course servings

CHICKEN CURRY WITH APPLES AND ONIONS

Any somewhat firm apple with lots of flavor is good here. This would be an ideal place to use **Black Oxfords** or **Blue Pearmains**, if you don't mind the skin. Ditto for **Rhode Island Greening**. Sweeter apples with rose or lychee scents, like **Pink Lady** and **Keepsake**, give a more exotic Indian flavor, and russet apples add nuttiness. **Arkansas Black** adds nice tannins and texture. **D'Arcy Spice** might be mind-blowing.

1 cup cider, hard or sweet

2-inch piece of ginger, peeled and roughly chopped

2 garlic cloves

1 teaspoon turmeric

2 teaspoons garam masala

½ teaspoon ground coriander

2 teaspoons vegetable oil or ghee

Salt and pepper

2 pounds boneless chicken thighs, extra fat trimmed off

1 large onion, chopped

2 shallots, chopped

1 cup chicken broth (or use more cider)

2 large apples, cored and diced

2–4 tablespoons butter

1 cup cilantro leaves, chopped

Apples are a natural in Indian cuisine. As proof, just think about the ingredients in garam masala, the classic North Indian spice mix: cinnamon, clove, cardamom, cumin, sometimes nutmeg. Sounds like apple pie to me. (The only outlier there is cumin, but its subtle presence in apple pie is not unwelcome.) Ginger, another curry standard, also plays well with apple. This curry is rich, fragrant, a touch sweet, and a real crowd-pleaser. Serve with basmati rice and chutney on the side, and a glass of cider. You can use boneless breasts or bone-in chicken parts here if you prefer. Twiddle with the cooking time appropriately.

1. Combine the cider, ginger, garlic, and spices in a blender or food processor and blend into a watery paste. Set aside.
2. Heat the oil in a large skillet over medium-high heat. Season the chicken with salt and pepper and add it to the pan. Cook until brown, about 3 minutes, then turn and brown the second side, another 2 minutes. Remove the chicken and set aside.
3. Add the onion and shallots and sauté until golden, about 5 minutes.
4. Add the cider paste and the chicken broth, stir, return the chicken to the pan, and cover. Reduce the heat to a simmer and cook, stirring occasionally, until the chicken is tender and cooked through, about 20 minutes.
5. Remove the cover, add the apple, increase the heat to medium-high, and cook, stirring, until the sauce has thickened and the apple is tender, 5 to 10 minutes.
6. Turn off the heat and stir in the butter. Sprinkle cilantro over the top and serve.

Makes 4 to 6 servings

PORK CARNITAS WITH APPLE SALSA

The salsa is best with one sweet and one tart apple, and two different colors add a little dazzle. **Yellow Transparent** and **Red Astrachan** in summer, **Ashmead's Kernel** and **Baldwin** in fall or winter, a **Granny Smith** and a **Fuji** together, etc. The dream team: **Ginger Gold** and **Pink Lady**, neither of which will brown much. **Orleans Reinette** makes a good soloist.

Carnitas is like Mexico's answer to pulled pork, but instead of slowly smoking your pork shoulder, you simmer it in its own fat in a cauldron over a fire until the liquid has evaporated and the pork ends up frying in the fat. The result is luscious shreds of meat with crispy edges. Traditionally, lime and orange are used to flavor the pork, but the Spanish certainly introduced the apple to Mexico in the 1500s, around the same time as the lime, so this recipe envisions an alternate history of Mexico in which the apple took hold. Cider vinegar gives it the tang, with hard and sweet cider providing flavorful cooking liquid. (Use all hard or all sweet if you prefer. If you don't have good cider vinegar for the salsa, use lime juice instead.) Fresh apples provide exactly the sweet/tart/crunchy elements that good salsa wants. In fact, their mix of sweet and tart isn't so different from a tomato's, so it's funny that we tend to think of one as savory food and one as dessert fare. I prefer parsley to cilantro in this salsa—it adds more of a fresh green taste—but both are good. *¡Buen provecho!*

For the Carnitas

¼ cup vegetable oil or lard

3 pounds boneless pork shoulder (aka Boston butt), **cut into chunks**

1 cup hard cider

1 cup sweet cider

¼ cup cider vinegar

2 teaspoons cumin

½ teaspoon cinnamon

10 garlic cloves

2 teaspoons salt

1 teaspoon pepper

For the Salsa

2 medium apples, cored and chunked

1 tablespoon cider vinegar

1 teaspoon salt

1 handful parsley or cilantro leaves

1 small chile, stemmed and seeded

1 garlic clove

For Serving

24 6-inch corn tortillas

1 medium onion, finely chopped

Lime wedges (optional)

Hot sauce (optional)

1. Start the carnitas: Combine the oil, pork, ciders, vinegar, and spices in a Dutch oven or a heavy pot with a lid. Bring to a boil over high heat, then reduce the flame to a simmer, cover, and cook for 1¼ hours.
2. Remove the lid and continue to simmer, stirring occasionally, until the liquid has evaporated, the meat is very tender, and it is frying and crisping slightly in the fat. The color will be pretty dark.
3. Make the salsa: Place all the ingredients in the bowl of a food processor and pulse quickly several times, until the apples are finely chopped but not pureed. Scrape into a serving bowl.
4. Heat the tortillas in a dry skillet or in the microwave.
5. To serve, place a line of meat (less than you think!) across a tortilla, and add chopped onions and salsa to taste. Serve accompanied by lime wedges and hot sauce.

Makes about 24 tacos. That's anywhere from 6 to 12 servings, depending on your crowd.

DUCK AND APPLE RISOTTO WITH BACON, SAGE, AND FOREST MUSHROOMS

I can't imagine any apple would be bad here, but ones that retain a touch of firmness are best, such as **Baldwin**, **Calville Blanc**, **Glockenapfel**, **Northern Spy**, and **Rhode Island Greening**. If it's a thick-skinned variety, you might peel it.

¼ pound bacon

Salt and pepper

2 large duck breasts (about 1 pound each)

2–4 cups fresh wild mushrooms

1 large onion, diced

2 cups arborio or (even better) **carnaroli rice**

1 cup dry cider

6 cups good chicken stock

2 large apples, cored and diced

15 fresh sage leaves, minced, plus more for garnish

Cider syrup for garnish (optional; see Resources)

The essence of fall. You can do this without the mushrooms, or without the duck, or without the bacon, and it's still very good. Together, with the cider and the sage, it will evoke intense sense memories of crispy days with yellow leaves dancing through the air. Chanterelles, porcini, and oyster mushrooms are all good here, and even portobellos will suffice. There's been a movement the past few years to make no-stir risotto; resist it. The slow release of starches by the rice is what gives you the creamy sauce, and there's no way to shortcut that. Besides, the half hour you spend stirring your risotto can be the highlight of your day if there's a bottle of cider beside you. One splash for the dish, one for the cook…

1. Preheat the oven to 250 degrees.
2. In a large skillet, fry the bacon over medium heat, turning once, until crisp. Set aside.
3. Liberally salt and pepper the duck breasts, rubbing the seasoning into the skin and flesh with your fingers. Add them to the skillet, skin side down, and cook until the skin is crispy, about 5 minutes. The duck breasts should release all the fat you need to cook the rest of this recipe. (That is one of the many beauties of duck breasts.) Turn the breasts over and cook the other side for an additional 2 minutes. Move the breasts to a baking sheet and keep warm in the oven.
4. Add the mushrooms and onion and cook, stirring occasionally, until softened, about 5 minutes.
5. Add the rice and cook, stirring, until the rice is glossy, about 2 minutes.
6. Add the cider and cook, stirring, until most of the liquid has evaporated.
7. Add ½ cup of the chicken stock and cook, stirring, until the liquid has almost entirely evaporated. Repeat with another ½ cup.
8. Add the apples and minced sage.
9. Add 1 cup of the stock and cook, stirring, until the liquid has almost evaporated. Repeat this step, stirring continuously, until the rice is tender but not soft, about 30 minutes. Some people like the rice to have a chalky core, but I prefer a chewy core.

10. When the rice is done, turn off the heat. Taste and adjust the seasoning.
11. Slice the duck breasts against the grain into ½-inch-thick slices. Serve the risotto in a wide, shallow bowl with a fan of duck slices on the side, and bacon crumbled over the top. Garnish with a sage sprig.
12. Optional: Drizzle one thin line of cider syrup across each serving.

Makes 6 to 8 servings

LAMB BRAISED IN APPLES AND SPICES

You actually want the apples to melt into the gravy here, so tenderness is prized, in both the flesh and the skin. **Yellow Transparents**, **Gravensteins**, **Cortlands**, and **Snow** are great. **Yellow Bellflower** would be spectacular.

1 boneless, butterflied leg of lamb (3–4 pounds)

Salt and pepper

2 tablespoons olive oil

1 cup beef broth

16 garlic cloves

6 large apples, cored and chopped

1 tablespoon cinnamon

1 tablespoon cumin

½ teaspoon coriander

2 pounds new potatoes

1 cup oil-cured black olives, preferably pitted

1 preserved lemon, seeds removed, chopped (optional)

Fresh parsley leaves, chopped, for garnish

Your traditional Moroccan-Kazakh-Irish stew. The apples cook down into a rich, brown, tangy gravy, punctuated by the bright salty bits.

1. Preheat the oven to 350 degrees. Season the lamb liberally with salt and pepper, rubbing the spices into the meat with your fingers.
2. Heat the olive oil in a Dutch oven over medium-high heat. Add the lamb and brown well, about 5 minutes. Turn and brown the other side, another few minutes. Turn off the heat.
3. Remove the lamb and set aside. Pour off the excess fat. Add the broth and scrape any browned bits off the sides of the pot. Return the lamb to the pot, cover, place it in the oven, and cook for 45 minutes.
4. Add the garlic, apples, spices, potatoes, olives, and lemon (if using), keep covered, and cook an additional 30 to 45 minutes, checking the lamb for doneness after half an hour. It's best medium-rare (pink but not purple). Continue cooking if necessary, adding a bit of liquid if it gets dry and sticky, and leaving the cover off if you want a thicker gravy. Adjust the seasoning if needed.
5. Slice the lamb and serve it on a platter, surrounded by the garlic and potatoes and topped with the gravy. Garnish with the parsley.

Makes 8 to 12 servings

SAUSAGE-APPLE-CHEDDAR POTPIE

Sweet apples work best, and firmness is not essential. **Summer Rambo** would be a fun surprise. **Golden Delicious** would be in its element. **Fuji** or **Gala** would be nice, **Pink Lady** even better. **Tolman Sweet** or **Jonathan** would give you a very old-timey pie, **Hudson's Golden Gem** a mod Pacific Northwest one, and **Grimes Golden** a distinctly southern spin.

A rustic pie with a cheese biscuit crust—serious comfort food, made even more comforting by the apples.

For the Crust

2 cups flour

2 teaspoons baking powder

½ teaspoon salt

4 tablespoons butter, cut into pieces

1 cup grated extra-sharp cheddar cheese

1 cup milk

For the Filling

1 pound pork sausage, bulk, or casings removed

1 medium onion, diced

2 large carrots, peeled and diced

2 celery stalks, diced, or ½ cup diced celery root

2 large apples, cored and diced

2 garlic cloves, minced

3 tablespoons flour

1 cup chicken or beef stock, or sweet cider

1 teaspoon dried, crumbled sage

Salt and pepper

1. To make the crust dough, combine the flour, baking powder, and salt in a food processor and pulse. Add butter and pulse until the mixture resembles coarse meal. Add the cheese and milk and pulse just until the dough forms and pulls away from the sides of the food processor. Set aside.
2. Preheat the oven to 400 degrees.
3. In a 9- or 10-inch cast-iron skillet, cook the sausage over medium-low heat until browned, breaking it up as it cooks. Remove the sausage and set aside.
4. Add the onion, carrots, and celery to the pork fat and cook over medium heat, stirring occasionally, 4 minutes.
5. Add the apples and garlic and cook, stirring occasionally, until the vegetables have softened, about 4 minutes.
6. Add the flour and stir until incorporated, about 1 minute.
7. Add the stock and sage and stir until a hot, bubbling gravy has formed, about 2 minutes. Return the sausage to the pan and stir. Turn off the heat. Taste and add salt and pepper as needed.
8. Drop the biscuit dough over the top in spoon-size balls. It's okay if it is uneven or if there are small gaps; it will spread out as it cooks.
9. Bake until the top is puffed and golden, about 30 minutes. Remove from the oven and let cool 10 to 15 minutes before serving.

Makes 8 servings

CIDER-BRAISED CABBAGE WITH APPLES AND FENNEL

This is a perfect spot for a russet, with its firmness and nutty taste. **Ashmead's Kernel**, **Zabergau Reinette**, **Roxbury Russet**, **Pomme Grise**, even **Golden Russet**. **Orleans Reinette** would be superb. If not using a russet, then any firm apple will do.

4 tablespoons butter, cut into pieces

1 teaspoon caraway seeds

½ head red cabbage, core removed and thinly sliced

½ fennel bulb, thinly sliced (reserve fronds)

1 large apple, cored, halved, and sliced

⅓ cup boiled cider (see Resources)

Salt and pepper

Let me introduce you to a wonderful product: Wood's Boiled Cider. Wood's Cider Mill, in Springfield, Vermont, has been making boiled cider since 1882. It's simply fresh apple cider reduced to a syrup with eight times the concentration of sweet cider (meaning it's both very sweet and very tart), and it gives you all sorts of culinary possibilities. It can intensify the flavor of pies and crisps, and it also makes a great braising liquid when you don't want too much water, as in this recipe. It's also pretty darn good over ice cream. (It can be easily ordered online, but if you don't have any, add sweet cider before the cabbage and fennel and reduce it by three-quarters before proceeding.) Serve this cabbage beside roast chicken or pork.

1. Melt 2 tablespoons of the butter in a large skillet with a lid over medium heat.
2. Add the caraway seeds and toast until fragrant, shaking the pan once or twice, about 30 seconds.
3. Add the cabbage and fennel, toss in the butter, and sauté 2 minutes.
4. Add the apples and boiled cider, cover the skillet, and cook, stirring occasionally, until the cabbage has softened but still retains some crunch, about 6 minutes.
5. Remove the lid, add the remaining butter, turn the heat to high, and sauté until any liquid has evaporated and the cabbage is coated in sauce, 2 to 3 minutes. Add salt and pepper to taste, and serve, garnished with a few fennel fronds.

Makes 4 to 6 servings

BRUSSELS SPROUTS WITH BACON AND APPLES

Russets such as **Pitmaston Pineapple**, **St. Edmund's Russet**, and **Golden Russet** complement the nuttiness of the Brussels sprouts, as does **Blenheim Orange**. Red-skinned apples add complementary color to the green Brussels sprouts.

Brussels sprouts are the best of the cabbages, bite-size and nutty, with a satisfying toothiness to them. They have a bad reputation because for years the ones on the market were terrible, and often terribly cooked. But good ones, braised in a sauce that supports their assertiveness, can be spectacular. This dish is a wonderful mix of sweet, savory, and salty, with lots of crunch. Serve with pork chops, venison, duck breast, or prime rib.

4 bacon slices

24 Brussels sprouts, halved

¼ cup boiled cider (see Resources) **or sweet cider**

1 medium-large apple, cored and diced

1 garlic clove, minced

½ cup heavy cream

Salt and pepper

1. Fry the bacon in a large pan with a lid over medium-low heat until crisp. Remove the bacon, chop, and set aside.
2. Add the Brussels sprouts and the boiled cider to the pan, stir to toss in the bacon fat, cover, and cook, stirring occasionally, about 6 minutes.
3. Add the apple and cook, covered, until the apples and Brussels sprouts are tender, another 4 to 5 minutes.
4. Uncover the pan, add the garlic, and cook until the garlic is golden, 1 to 2 minutes.
5. Add the cream, raise the heat to high, and cook, stirring constantly, until the sauce is thick and bubbling, just a minute or two. Turn off the heat and add salt and pepper to taste and half the bacon. Toss.
6. Serve immediately with the remaining bacon over the top.

Makes 4 to 6 servings

MAPLE APPLESAUCE

Tender, thin-skinned apples like **Yellow Transparent** and **Gravenstein** are superb. **Grimes Golden** and **York** are famous saucers. **Cortland**, **Snow**, and **Mac** are also good. **Golden Delicious** makes fantastic sauce, especially tree-ripened ones from a good southern or western orchard. A mix of several different varieties always makes the most compelling sauce.

I've purposely kept this recipe very loosey-goosey, because the point is that you don't need to sweat the details or the quantities. Just throw apples in a pot with some liquid to keep them from burning, cook until soft, and puree. It takes just a few minutes and is a great accompaniment to any roast meat, to potato pancakes, or atop ice cream for dessert. Most applesauce recipes call for peeling the apples or pressing the sauce through a food mill to remove the skins, but to me, the skins are the point; they provide most of the flavor and color, and a food processor makes quick work of them. True, you don't get an ultra-smooth sauce this way, but I actually prefer the texture they provide, and the color is a big bonus. A little butter at the end gives it a velvety feel and a luscious flavor.

A bagful of apples, cored and quartered

At least a cup of sweet cider or water (or lemonade, even)

A drizzle of maple syrup

A few shakes of spices to your liking (the tiniest touch of cayenne can be quite nice)

A few tablespoons of butter

1. Place the apples and cider in a pot, cover, and cook over medium-low heat, stirring to make sure the apples don't stick to the bottom and burn. Add more liquid if you need to.
2. After about 10 to 15 minutes the apples will soften and begin to give up their juice. Add the maple syrup and spices and continue cooking, stirring occasionally, until the apples are fully soft and mushy, another 10 minutes or so.
3. Add the butter, stir, and puree everything in a food processor, being careful not to scald yourself. (An immersion blender works well, too.) Taste and add more sweetness or spices if desired. Serve hot or cold.

APPLE-LIME CUSTARD TART

I like to make this with **Yellow Transparents** off the tree in my backyard as soon as they are ready in August. Their thin skin and light, tart flavor are a good fit. But this is also a perfect opportunity to show off apples with pronounced citrus or pineapple flavors, such as **Belle de Boskoop**, **Ananas Reinette**, **Orleans Reinette**, **Cox's Orange Pippin**, **Pinova**, and **Karmijn de Sonnaville**. I dream about making this with **Westfield Seek-No-Further**.

For the Crust

2½ cups pecans

4 tablespoons butter

4 tablespoons sugar

¼ teaspoon salt

For the Filling

4 large apples, cored and quartered

3 eggs

1 cup sugar

Zest and juice from ½ lime

Optional, for Serving

Whipped cream and candied citrus peel

I was once looking for a way to use apples fast, with a minimum of prep work, and I wanted to preserve as much of the fresh, flowery flavor of the peels as possible, and I came up with this oddity, which has since become a staple in our household, where it does double duty as breakfast and dessert. It's essentially an apple key lime pie. Too often, apple desserts all go in the same direction: heavy on the brown sugar, caramelized apple, and cinnamon. This is much lighter and zingier. It is also gluten-free. Serve it as is or with whipped cream and a touch of candied citrus peel. If you have a premade nut or graham-cracker crust on hand, it can be made in minutes.

1. Preheat the oven to 375 degrees.
2. Make the crust: Blend the pecans, butter, sugar, and salt in the bowl of a food processor until fine.
3. Grease a 9-inch pie pan. With your fingers, press the nut mixture into the bottom and sides of the pan.
4. Bake the crust until lightly browned, about 10 minutes. Remove from the oven.
5. Make the filling: Combine the apples, eggs, sugar, and lime zest and juice in the bowl of a food processor. Pulse until everything is thoroughly mixed and the peels have been reduced to flecks.
6. Pour the filling into the crust and bake for 25 to 30 minutes, until the custard has set but still quivers. Remove from the oven, let cool, and refrigerate. Serve cold.

Makes 8 servings

APPLES AND ORANGES GALETTE

Flavorful and tart. This is the moment **Belle de Boskoop** has been waiting for all its life. Also a good spot for russets: **Ashmead's Kernel**, **D'Arcy Spice**, or **Roxbury Russet**. **Northern Spy** is just about perfect, as usual. Those will all stay crisp—possibly too crisp for some. If you prefer more melting fruit, try **Bramley's Seedling**, **Smokehouse**, **McIntosh**, or **Jonathan**.

For the Crust

1⅓ cups all-purpose flour

2 tablespoons confectioners' sugar

8 tablespoons (1 stick) **cold, salted butter, cut into pieces**

1 egg, cold

For the Filling

½ cup orange marmalade, other preserves, lemon curd, or smooth peanut butter

2 large apples, cored and halved

2 tablespoons butter

1 tablespoon brown sugar

¼ cup bourbon-soaked raisins (see page 299), **drained**

½ cup pecans, chopped

Galettes are rustic fruit tarts that are supposed to look charmingly unpolished, which plays well to my kitchen strengths. To hold together, they need to have more strength than a pie crust; hence the egg in the dough. Although the main recipe calls for orange marmalade, there are limitless potential variations, including peanut butter, which exalts the simple lunchtime concept of apple slices with peanut butter.

1. Make the crust: Combine the flour and confectioners' sugar in a food processor and pulse for a second or two. Add the butter and pulse a few times until it resembles coarse meal. Add the egg and pulse once or twice until just combined.
2. Dump the dough onto a floured surface and gather it into a ball. Wrap the ball in plastic wrap, press it down into a disk, and refrigerate for at least 30 minutes.
3. Preheat the oven to 400 degrees.
4. Unwrap the dough and place it on a well-floured surface. Using a rolling pin, roll the disk from the center outward until it forms a circle of at least 12 inches. Perfect roundness is not required, or even desired.
5. Spread the marmalade thinly over the surface of each tart, leaving a 1- to 2-inch rim plain.
6. Slice the apples, but keep the halves together. Place them on top of the tart, skin side up, and nudge them to fan them out slightly.
7. Fold the edges of the pastry over the edges of the filling to create a rim. Start with a corner of the pastry, then do a side, then a corner, and so on, each overlapping the corner of the previous fold. You should end up with a shape a bit like an uneven Stop sign. The tart should stay open in the middle.
8. Melt the butter in a skillet over medium heat. Add the brown sugar and cook, stirring constantly, for one minute, until bubbly. Brush the exposed rim of the tart with a bit of the sugar-butter. Pour the rest over the apples. Sprinkle the raisins and pecans over the top.

9. Bake until the tart rim is dark brown, about 30 minutes. Don't burn it, but don't take it out too early (when it is just lightly golden), or it won't have enough strength later.
10. Allow to cool completely before serving, at least 20 minutes.

Makes 8 servings

NUTTY APPLE CRISP FOR A CROWD

You want tart, citrusy varieties, such as **Belle de Boskoop**, **Karmijn de Sonnaville**, and **Ribston Pippin**. Russets, especially **Orleans Reinette** and **Zabergau Reinette**, enhance the nutty factor, as does **Blenheim Orange**. As with pie, you get the most intriguing crisps if you mix at least three kinds of apples together.

8 large apples, cored and diced

Zest and juice of ½ lemon

½ cup granulated sugar

1½ cups all-purpose flour

1 cup brown sugar

2 cups pecans

2 teaspoons cinnamon

8 tablespoons (1 stick) **cold butter, cut into pieces**

Why make small apple crisps when you can make big ones for no extra work? Years ago, I gave up on oats in my crisp toppings—always dry and chewy—and switched to nuts, which make the topping richer, crunchier, and, well, nuttier. Pecans are my nut of choice. (If Europe had had pecans and America walnuts, instead of the other way around, walnuts would still be suspect.) Sliced almonds are also excellent—but add them after you've processed the rest of the topping, so they can act as crunchy little wafers. I've also decided that crisps are better when the apples are chopped pretty small—it keeps the topping even, which helps it cook more uniformly.

1. Preheat the oven to 375 degrees.
2. In a large bowl, toss the apples with the lemon and granulated sugar. Spread the apples into a 13 x 9 x 2-inch baking pan.
3. Add the remaining ingredients to the bowl of a food processor and pulse until they resemble coarse meal, just a few seconds. Spread the topping evenly over the apples and bake 45 to 50 minutes, until the topping is mahogany brown.
4. Remove from the oven and let cool. Serve warm or at room temperature.

Makes 12 servings

LEMON-CIDER SORBET

Marialisa Calta and her husband, Dirk van Susteren, make cider from their own apples, which tends to be tarter and more aromatic than commercial stuff; look for the freshest, tangiest cider you can find (unpasteurized if possible), or spike it with extra lime or lemon juice.

My neighbor Marialisa Calta, author of *Barbarians at the Plate* and coauthor of *River Run Cookbook*, is a legend in Vermont dining circles. I have been charmed by her Lemon-Cider Sorbet more than once, finding it the ideal way to cap an evening. After much cajoling, she relented and gave me her recipe. The secret is the lemon and the booze, which make it light, refreshing, and uplifting.

½ cup water

½ cup sugar

1 lemon, possibly more

2½–3 cups sweet cider

2–4 tablespoons Calvados, apple brandy, or bourbon

1. Pour the water into a small saucepan. Add the sugar. Peel one lemon in large chunks of peel, avoiding the white (bitter) pith, and add the peel to the pot. (Don't discard that peeled lemon.) Bring to a boil, reduce heat, and simmer for 5 minutes. Allow to cool to room temperature. You can do this part several days ahead of time. Cover and refrigerate until ready to use. Leaving the lemon peel in while the syrup cools infuses the syrup with lemon flavor. Keeping the pieces large makes it easy to remove them when the time comes.
2. Remove the lemon peel from the sugar syrup and discard. Combine the syrup, cider, and 2 tablespoons of the booze in a bowl. Squeeze the lemon and add the juice to the bowl. Stir, cover, and refrigerate until thoroughly chilled, at least 1 hour and up to 24. Taste. If it needs more lemon juice or booze, add some.
3. Pour into the bowl of your ice cream maker and proceed according to the manufacturer's directions. Alternatively, freeze in a baking pan overnight. About 3 hours before serving, break the frozen concoction into pieces and puree in a food processor until smooth, adding a few spoonfuls of fresh cider if you need to so it will smush up. Freeze again, covered, until ready to serve.

Makes about 5 cups

TARTE TINTIN

With so few ingredients, the apples really need to carry this. Tarte Tatin is **Calville Blanc**'s whole raison d'être, but any firm, tart apple will do, and a mix is even better **GoldRush** is fantastic, as are **Esopus Spitzenberg**, **Rhode Island Greening**, **Roxbury Russet**, **Newtown Pippin**, **Winesap**, **Mutsu**, and **Granny Smith**.

8 tablespoons (1 stick) **butter**

1 cup sugar

6–8 large apples, cored, halved, and sliced

1 14-ounce package puff pastry

This is my kind of apple pie: no rolling, no fussing, no pastry making, and shockingly few dishes to clean up afterward. I always find the bottom crust on apple pie disappointingly mushy anyway, so eliminating it seems like an inspired solution. The trade-off is that when you serve it, it doesn't hold together so well, but who cares; it isn't going to last that long anyway. Essentially, this is a tarte Tatin that you don't flip (so you don't have to be too worried about what your apple slices look like underneath)—although you can flip it if you want. A lot of tarte Tatin recipes call for puff pastry, but I find that the pastry gets too soggy and condensed once flipped; with this one, the pastry stays high and puffy and crisp and crackly. Named in honor of the Belgian boy reporter who was forever getting flipped when he didn't want to be.

1. Preheat the oven to 400 degrees.
2. Melt the butter in a 9- or 10-inch cast-iron skillet over high heat. Add the sugar and cook, stirring regularly, until the caramel turns golden, about 5 minutes.
3. Add the apples and cook, stirring frequently, until they have absorbed the caramel and everything has turned dark amber, 8 to 10 minutes. Turn off the heat.
4. On a lightly floured surface, roll out the puff pastry and nestle it over the apples in the cast-iron skillet, tucking it down around the sides if possible.
5. Bake about 25 minutes, until the top has turned brown and puffy. Let cool completely, so the insides can gel, before serving.

Makes 8 servings

BAKED APPLES WITH BOURBON-SOAKED RAISINS

Recipes for baked apples always tell you to go big, but a large baked apple is too much for many people, so consider medium instead. The "Baker's Buddy," **Rome Beauty**, was made to order here. Definitely go for an apple that will hold its shape and not turn to mush. **Wolf Rivers** are famous for baked apples, and **Black Oxfords** and **Blue Pearmains** are classics. Henry Ward Beecher singled out the **Tolman Sweet** as the best apple for baking in his sermon on apple pie. The modern classic, for baked apples, is the **Pink Lady**, which Russ Parsons of the *Los Angeles Times* tagged back in 1998 as "The Apple of the Future." Pink Ladies are sweet, spicy, complex, and firm, and their skin stays pink after baking, adding measurably to the wow factor.

1 cup pecans

¾ cup bourbon-soaked raisins (see right)

Zest and juice of 1 lemon

6 tablespoons butter, softened

½ cup brown sugar

Pinch of nutmeg

6 medium-large apples

⅔ cup sweet cider

The vanilla-scented flavor of bourbon goes beautifully with apples and pecans; the question is how to get it in there. My favorite method is to let raisins grab the bourbon ahead of time; they will slowly exude its essence into the apples as you eat them. The raisins need to be made at least a day ahead of time to really soak up the bourbon, but you can make a big batch and keep it for months in your pantry. They will come in handy all the time. Serve with vanilla ice cream.

1. Preheat the oven to 375 degrees.
2. Toast the pecans by placing them on a sheet in a toaster oven and toasting on a medium-low setting. They should be dark brown and crackling. Be careful not to burn them. Set aside to cool.
3. Chop the pecans. Combine everything except the apples and cider in a bowl and mix.
4. Core the apples down to the seed layer, using a paring knife, leaving the bottom of the core in place to hold the filling in. You should have a bowl-shaped cavity in the top of each apple when you are done. Place the apples in a baking pan.
5. Fill the cavity of each apple completely with the butter-raisin mix. If you have extra filling, save it for step 7. Pour the cider in the pan around the apples.
6. Cover the baking pan with foil and bake for 40 minutes.
7. Remove the foil, add any extra filling to the apple cavities, and bake an additional 15 to 20 minutes. Some apples soften much more quickly than others. When a fork or knife pushes easily into the apples, they are ready. Remove from the oven, cool for 5 minutes, and serve in low bowls with the pan sauce poured over the top.

Makes 6 servings

Bourbon-Soaked Raisins

Place 1 cup of raisins in a glass jar. Fill the jar with 3/4 cup bourbon (or applejack or Calvados). Seal the jar and store at room temperature. The raisins will start absorbing the bourbon within an hour, but they need to sit overnight for full power. You can keep reusing the bourbon for more batches, or you can make a killer Manhattan with it.

APPLE PIE SQUARES

Tart, firm apples all the way: **Calville Blanc**, **Bramley's Seedling**, **Newtown Pippin**, **Rhode Island Greening**, **Mutsu**. **Glockenapfel** or **Belle de Boskoop** if you can find them. Use a few different varieties if you can.

All the goodness of a pie, in convenient form. By increasing the ratio of crust to filling, and making the crust sturdier with egg and sugar, this dessert becomes eminently transportable. I've also pushed the spices into cinnamon-roll territory.

For the Crust

2¾ cups all-purpose flour, plus 2 tablespoons

½ teaspoon salt

1½ cups sugar

½ pound (2 sticks) **butter, cut into pieces**

1 egg

½ cup milk, plus 2 tablespoons for brushing

For the Filling

5 large apples

2 teaspoons cinnamon

¼ teaspoon ground cloves (optional)

¼ teaspoon cardamom (optional)

Pinch of nutmeg (optional)

1. Put 2¾ cups flour and the salt in the bowl of a food processor. Add ½ cup of the sugar. Pulse briefly to mix. Add the butter and pulse until the mixture resembles coarse meal. Add the egg and ½ cup milk and pulse to form a dough, stopping as soon as the mixture begins to pull away from the sides of the bowl.
2. Gather the dough into a ball and divide it in two. Wrap the two halves in plastic wrap and refrigerate for at least 1 hour.
3. When ready, preheat the oven to 400 degrees.
4. Grease a 9 x 9-inch baking pan.
5. Core and dice the apples. In a large bowl, toss them with the spices, 2 tablespoons of flour, and almost 1 cup of the sugar, reserving 2 tablespoons to sprinkle over the top at the end.
6. Roll out one ball of dough on a floured hard surface until it conforms to the baking pan. Lay it in the pan. (The best way to move the dough around is to fold it in quarters and then unfold it in the pan.)
7. Top the dough with the apple mixture, smoothing the apples into a fairly flat layer.
8. Top the apples with the remaining layer of dough. Brush the top with the 2 tablespoons milk, then sprinkle the remaining 2 tablespoons of sugar over the top. Bake for 40 to 45 minutes, until golden brown on top. Let it cool completely before cutting it into squares, or it won't hold together.

Makes about 16 squares

KATE McDERMOTT'S ART OF THE APPLE PIE

Kate favors a mix of sweet and tart apples, such as **Newtown Pippin**, **Golden Russet**, **Esopus Spitzenberg**, and **Bramley's Seedling**. The greatest single-variety pie apple in the world, she says, is **Gravenstein**, intensely sweet and tart.

For the Crust

2½ cups unbleached all-purpose flour, plus more for rolling the dough

½ teaspoon salt

8 tablespoons (1 stick) **butter, cut into pieces**

8 tablespoons leaf lard

6–8 tablespoons ice water, plus more if needed

For the Filling

10 cups quartered, cored, and chopped apples

½ cup flour

½ cup sugar

½ teaspoon salt

1 teaspoon cinnamon

2 small gratings nutmeg

½ teaspoon allspice

1 tablespoon quality apple cider vinegar (not industrial)

1–2 teaspoons butter, chopped into little pieces

1 egg white mixed with 2 tablespoons water

1 tablespoon sugar

My friend Kate McDermott makes the most perfect apple pies I know, and she has the bling to prove it: a sheaf of blue ribbons from the Clallam County Fair, not to mention first prize in Seattle's Pie and Poetry Slam. About a decade ago, Kate set out on a two-and-a-half-year odyssey to develop the perfect pie crust, mixing all sorts of different flours, fats, and techniques, and then followed that up with the perfect apple pie recipe, auditioning seemingly every apple variety in Washington State. This the child of that labor. Kate (like many others) believes that leaf lard, the fat from around the kidneys of a pig, is the key to flaky, flavorful crusts. She also favors King Arthur flour and Kerrygold Irish butter.

1. Combine the flour, salt, butter, and lard in a large bowl.
2. With clean hands, blend the mixture together until it looks like coarse meal with some pea- to almond-size lumps in it. The lumps make flaky pies.
3. Sprinkle the ice water over the mixture and stir lightly with a fork. Squeeze a handful of dough together. If it doesn't hold together, mix in a bit more water.
4. Pull the dough together and compress it into a big ball. Cut the ball in half and shape it into two chubby disks. Wrap each disk tightly in plastic wrap and chill in the fridge for about an hour.
5. When you are ready to roll out the dough, preheat the oven to 425 degrees. Take the dough out of the fridge. If it feels hard and solid, let it rest at room temperature until it feels a bit more pliable. Place a handful of flour onto a pastry cloth, parchment sheet, or flat dry surface. Unwrap one disk and place on the flour. Turn so both sides are covered with flour. Now take your rolling pin and thump the dough a few times on each side to wake it up.
6. Sprinkle more flour on the top of the dough if needed to keep the rolling pin from sticking and roll the dough out from the center in all directions. When it is an inch or so larger than your pie pan, fold the dough over the top of the rolling pin and lay it in the pie pan carefully. Don't worry if the crust needs to be patched together; just paint a little water where it needs to be patched and "glue" on the patch piece.
7. Place the apples, flour, sugar, salt, spices, and vinegar in a large bowl and mix lightly until most of the surfaces are covered. Pour the apple mixture into the pie crust, mounding it high, and dot with the butter.

8. Roll out the second disk of dough and place it on top. Crimp the edges with a fork. Cut a few vent holes in the top crust, paint with the egg-white wash, and sprinkle the sugar over the top.
9. Place the pie in the center of the oven and bake for 20 minutes.
10. Reduce the heat to 375 degrees and bake for 40 minutes longer.
11. Let the pie cool for at least 1 hour before eating—if you can!

Makes 8 to 12 servings

GLOSSARY

Beach, Spencer Ambrose. Horticulturalist at the New York Agricultural Experiment Station in Geneva and author of *The Apples of New York* (1905), still the most important apple reference.

Bloom. A natural, powdery wax exuded by apples to waterproof their skin. All apples protect their skin with waxes, but some apples produce waxes with crystalline structures that scatter light in particular ways, producing a colorful "bloom" that can be rubbed off with the hand.

Bunker, John. Founder of Fedco Trees in Maine, discoverer of many lost varieties, and author of the book *Not Far from the Tree: A Brief History of the Apples and the Orchards of Palermo, Maine, 1804–2004.*

Bunyard, Edward. Scion of England's greatest nursery family, founder of the *Journal of Pomology*, tireless proselytizer for fruit trees, and author of the 1929 classic *The Anatomy of Dessert*. Bunyard's witty prose and keen gastronomic observations are as spot-on today as they must have been eighty-five years ago.

Burford, Tom. Legendary, seventh-generation Virginia horticulturalist, consultant to many heirloom orchards, including the restored one at Monticello, and author of *Apples of North America* (2013).

Coxe, William. New Jersey pomologist and author of the first book on American apples, *A View of the Cultivation of Fruit Trees, and the Management of Orchards and Cider* (1817). Coxe provides the best window into the role of apples in early America.

Downing, Andrew Jackson. American landscape designer and author, with his brother Charles, of the 1845 *Fruits and Fruit Trees of America*, which remained the States' most important work on apples until *The Apples of New York* came along.

Eye. Nickname for the apple calyx, the remnants of the flower sepal, which protrude from the bottom center of the fruit. Some apples have open eyes, others closed.

Goodband, Zeke. Manager of Scott Farm, one of the premier producers of heirloom apples and a major factor in the current heirloom revival.

Graft. The technique of inserting a scion, or shoot, from an apple tree into the cleft of a separate rootstock or branch of another tree.

Hogg, Robert. Towering figure in British pomology, and author of, yes, *British Pomology* (1851). Hogg was the seminal force behind the founding of the British Pomological Society and a great celebrant of British apple varieties, which at the time were being driven toward extinction by American imports.

Lenticels. Pores on the skin of an apple through which the fruit breathes. These show up as dots or specks and are often a defining feature.

Oblate. "Depressed at the poles"; used to describe an apple with a slightly flattened shape, as if it were gently stepped upon.

Pearmain. An ancient word that turns up in the names of dozens of apples, going back hundreds of years. Different sources give different explanations of the word, none of them satisfactory. Some claim "pearmain" refers to apples with a pearlike flavor, others that it indicates pear-shaped apples. Yet the apples named "pearmain" do not all fit either possibility, and don't seem to have anything else in common, either. Other sources believe that the original "Parman" referred to an apple from Parma, Italy, or that it stems from the Old French *parmaindre*, "to remain," in other words, a good keeping apple that would hang around through the winter. That last option can be accurately applied to most pearmains, and it's my best guess.

Pippin. Old British word for "seedling." A pip is a seed, and a pippin is a sprout that grew from seed. "Cox's Orange Pippin" indicates a tree grown from seed (rather than by grafting from a known variety) by Cox.

Reinette. A common epithet for many French apples, such as Ananas Reinette, Orleans Reinette, and Reine des Reinettes. English speakers tend to assume that it must mean "little queen" or "princess," for obvious reasons, but there is no such word in French. Another common fallacy is that *reinette* is the Gallic equivalent of "pippin," and refers to a seedling—an idea that may have stemmed from the fact that the English King of the Pippins and the French Reine des Reinettes may be one and the same apple. People also like to mention the French colloquialism *rainet*, which means "little frog," but why so many apples would be named after tree frogs is beyond me. The best theory I've heard is that *reinette* is actually a corruption of the Latin *renatus*, "born anew," and refers to a sport, a mutant branch that produces its own genetically distinct variety (see below). Whatever the case, most of the reinettes tend to be hard, tart, late-ripening russet apples with complex flavors, so it's a good word to look for.

Ribs. All apples have a five-sided geometry, which goes back to the original five-petaled flower. Each fruit is actually a melding of five fleshy lobes, each corresponding to a single seed-containing ovary. (When one of the flower's ovaries fails to get pollinated, it won't develop, and that part of the fruit will be folded in on itself.) The ribs are the five points of the pentagon. Some apples are better than others at rounding off these ribs. The Api Étoile (page 236) has the most prominent ribs of any apple (though I've discovered a few wild ones that will give it a run for its money).

Russet. A rough, sandpapery covering on an apple skin, almost always brown, tan, or green. Americans, who have been taught to expect waxy, shiny apples, tend to avoid russets, yet almost without exception, russet apples have intense flavors (often nutty) and crunchy textures, and they are favored by the English. Perhaps the genes that trigger russeting also intensify flavor, but another possibility exists: russeted skins are more permeable to water, so russet apples may be constantly evaporating water weight while hanging on the tree, slowly concentrating like a good sherry.

Scion. A new shoot that is cut from the parent tree and grafted onto a separate rootstock to produce a tree that will be an exact clone of the parent. This is how all apple varieties are propagated.

Shand, Morton. Witty British food and wine writer of the 1920s (and the grandfather of Camilla Parker Bowles).

Shoulders. The highest ridges of an apple, which rise from the bowl of the stem cavity before curving down the sides.

Sport. Occasionally a single branch of an apple tree will suffer a genetic mutation, causing it to produce apples slightly different from the rest of the tree. These "sports" can then be grafted to produce entire trees carrying their genetics. Some varieties, such as Winesap and Red Delicious, have many popular sports, often chosen because they are redder (though, sadly, usually less tasty) than the original.

Sprightly. Apple jargon for a lively, tart taste.

Subacid. Apple jargon meaning "a little bit tart."

Suture Mark. A slightly indented line running from the stem to the eye of certain apples, as if the skin were stitched on.

Wood, Steve. The Prometheus of American cider. Back in the 1970s, Wood saw that his small New Hampshire orchard was not going to be able to compete with the behemoths of Washington State and China, so he changed direction, traveling to Europe to bring back scion wood for true cider apples at a time when few people in America knew what that meant. Not only was his Farnum Hill cider one of the first quality American hard ciders on the market, but his gift of scion wood has been the fire that kindled the creative flames of the new generation of cider makers.

RESOURCES

Mail-Order Apples

Alyson's Orchard
WALPOLE, NH
603-756-9800
alysonsorchard.com
Heirloom-sampler and Honeycrisp gift boxes, shipped September to November.

Stemilt Growers
WENATCHEE, WA
800-315-2306
bountifulfruit.com
One of the few sources for SweeTango apples, as well as Honeycrisp, Pink Lady, Gala, Pinova, Jonagold, and others.

Gray Wolf Plantation
NEW OXFORD, PA
717-624-7204
graywolfplantation.com
Specializes in mail-order gift boxes of pre–Civil War apple varieties.

Harmony Orchards
TIETON, WA
harmonyorchards.com
Top producer of organic apples in Washington State's Yakima Valley. Fantastic apple gift boxes during the holiday season.

Tree-Mendus Fruit
EAU CLAIRE, MI
877-863-3276
treemendus-fruit.com
Hundreds of varieties of apples in season.

Mail-Order Apple Trees

Century Farm Orchards
ALTAMAHAW, NC
336-349-5709
centuryfarmorchards.com
Has about four hundred apple varieties in its library, with an emphasis on southern and disease-resistant ones.

Cummins Nursery
ITHACA, NY
607-227-6147
cumminsnursery.com
Legendary Finger Lakes nursery offering forty to fifty varieties of apples on disease-resistant rootstock.

Fedco Trees
WATERVILLE, ME
207-426-9900
fedcoseeds.com/trees
John Bunker offers more than fifty varieties of apple trees, from classic heirlooms like Black Oxford and Kingston Black to modern masterpieces like Honeycrisp and Sweet Sixteen.

Greenmantle Nursery
GARBERVILLE, CA
707-986-7504
greenmantlenursery.com
This is the fine nursery operating on the old stomping grounds of Albert Etter, the eccentric Calfiornia plant breeder. Greenmatle's 250 apple varieties include many originally created by Etter.

Stark Bro's
LOUISIANA, MO
800-325-4180
starkbros.com
The legendary creator of Red Delicious, Golden Delicious, and many other apples.

Trees of Antiquity
PASO ROBLES, CA
805-467-9909
treesofantiquity.com
Huge selection of organic apple trees, including many very unusual varieties.

Vintage Virginia Apples
NORTH GARDEN, VA
434-297-2326
vintagevirginiaapples.com
More than two hundred varieties of organic apples in the Charlottesville region.

Boiled Cider

Wood's Cider Mill
SPRINGFIELD, VT
802-263-5547
woodscidermill.com
Syrupy-thick boiled cider; indispensible in recipes of all kinds.

Hard Cider

Aaron Burr Cider
WURTSBORO, NY
845-468-5867
aaronburrcider.com
The closest you can get to the home ciders of nineteenth-century America. Aaron Burr tracks down wild and abandoned apples in New York's Shawangunk Valley and ferments them using indigenous yeasts. The results are as wide-ranging as each year's weather, but always enigmatic and enlightening.

Albemarle CiderWorks
NORTH GARDEN, VA
434-297-2326
albemarleciderworks.com
Some of the country's best single-varietal ciders, including Virginia Hewes Crab and Old Virginia Winesap, and beautifully balanced blends. Also a handsome, light-filled tasting room.

Annandale Cidery
ANNANDALE-ON-HUDSON, NY
845-758-6338
mporchards.com
The superb, complex, and very hard to find single-varietal ciders from this historic Hudson River orchard include a heady Hewes Crab, a pure Newtown Pippin, and a staggeringly flavorful and just bearably tart Wickson. Well worth seeking out.

Eden Ice Cider
WEST CHARLESTON, VT
802-334-1808
edenicecider.com
Eden makes America's top ice cider, a tart, syrupy indulgence, and they are in the forefront of the next cider evolution: aperitifs. Their Orleans Herbal is dry cider infused with basil and anise hyssop; it is the elixir of the gods. Their Orleans Bitter is like an American Campari, with a red tinge from currants and a seriously bitter bite from gentian, angelica, and dandelion.

Farnum Hill Ciders
LEBANON, NH
603-448-1511
povertylaneorchards.com
Steve Wood has the best cider orchards in America (you should see the crates of Kingston Black coming in) and makes some of the most distinctive cider. Look for the Dooryard blends, which always push the tannin envelope.

Foggy Ridge Cider
DUGSPUR, VA
276-398-2337
foggyridgecider.com
Made in Virginia's Blue Ridge Mountains, Foggy Ridge's Serious Cider shows off Dabinett's lip-smacking tannins, and their Handmade is 100 percent Newtown Pippin, a drink Virginia's Founding Fathers would have recognized. Look for an increasing percentage of Harrison in their First Fruit, a blend of early-season varieties, as their Harrison plantings mature.

Tieton Ciderworks
TIETON, WA
509-673-2880
tietonciderworks.com
Six different ciders, including dry-hopped.

Uncle John's Cider Mill
ST. JOHN'S, MI
888-56-CIDER
fruithousewinery.com
Nicely tannic Rosé, Melded, Baldwin, and Russet ciders. Also cider in cans.

West County Cider
COLRAIN, MA
413-624-3481
westcountycider.com
Pioneers in the cider revival, West County makes what you might call purist ciders: tart, low-alcohol quaffers in the unadorned style. Their single-varietal Reine de Pomme is both elegant and funky, and their pink-tinged Redfield is gorgeous.

Festivals and Events

Capitol Cider
SEATTLE, WA
206-397-3564
seattleciderbar.com
Not a festival, but the first cider-focused bar and restaurant in America, the advent of which is a celebration in itself. With at least fifty ciders in its library and another ten to twenty by the glass, plus a full menu of cider-friendly fare, Capitol Cider is one of a kind.

Charlevoix Apple Festival
CHARLEVOIX, MI
Mid-October
charlevoix.org
For thirty-five years, Charlevoix has been showcasing the great apples Michigan can grow. This three-day fest includes at least thirty varieties for tasting, plus plenty of kielbasa and pasties for those who can't fill up on apples alone.

CiderDays
FRANKLIN COUNTY, MA
Early November
ciderdays.org
The highlight of my apple year, and the Mecca for every serious cider enthusiast in America. CiderDays draws home and professional cider makers, as well as plain-old enthusiasts, from around the country for two days of cider madness. There are cider-making workshops and apple tastings, cooking demos, an amateur cider contest featuring some of the oddest beverages you will ever taste, and a cider salon featuring the largest selection of ciders in the world, including top producers from Asturias, Normandy, and England. All that and an apple pancake breakfast, too.

Cider Summit
SEATTLE, CHICAGO, AND PORTLAND (OR)
Dates vary
cidersummitnw.com
Pourings from numerous cider producers in the Northwest, as well as Europe.

Cider Week
NEW YORK CITY AND HUDSON VALLEY, NY
Mid-October
ciderweekny.com
A ten-day bash of tastings, dinners, cider crawls, and workshops. Many of the city's bars and restaurants expand their cider menus to showcase the many faces of the Beautiful Fruit. Other cider weeks have now taken off in Virginia, Seattle, the Finger Lakes, and Michigan.

Gravenstein Apple Fair
SEBASTOPOL, CA
Mid-August
gravensteinapplefair.com
A celebration of the Gravenstein apple, once ubiquitous in Sonoma County. Lots of music, food, cider, and pie-eating contests.

Great Lakes Cider and Perry Festival
ST. JOHNS, MI
Early September
greatlakescider.com
Dozens of Midwest ciders to taste, along with live music and good eats.

Great Maine Apple Day
UNITY, ME
Late October
mofga.org
One of the greatest displays of heirloom apples you'll ever see. Pie contests, cider pressing, pruning workshops, and John Bunker tries to identify mystery apples brought from across Maine and beyond.

Heirloom Apple Day
SCOTT FARM, DUMMERSTON, VT
Mid-October
scottfarmvermont.com
Apple guru Zeke Goodband introduces visitors to some of the ninety varieties he grows at Scott Farm, one of the handsomest spots on earth. (It's where *The Cider House Rules* was filmed.) Fresh apples, pies, and lots of lore.

Heirloom Weekend
KIYOKAWA FAMILY ORCHARDS
PARKDALE, OR
Late October
mthoodfruit.com
Taste one hundred varieties of apples, plus pears and hard cider.

Monticello Apple Tasting
MONTICELLO, VA
Late October
monticello.org
A wildly popular two-hour program on the site of Thomas Jefferson's orchard, led by Tom Burford, aka Professor Apple. Participants taste and learn the history of a variety of classic southern apples, and then it all comes down to the big vote for best-tasting apple.

Vintage Virginia Apples Harvest Festival
NORTH GARDEN, VA
Early November
albemarleciderworks.com
Tons of apple varieties for tasting, cider pressing, hard cider tasting room, and hay rides.

ACKNOWLEDGMENTS

I am no apple expert. I'm more of an apple stalker, with a fixation bordering on the unhealthy. My son no longer likes to ride in the car with me during autumn. All this stalking has built up a certain base of apple knowledge, but it pales in comparison to the pros, without whom this book would have gone nowhere. In a sense, this book is simply a bushel of low-hanging fruit collected from their lifelong labor.

It all started with my neighbor Terry Bradshaw, who has been tending the University of Vermont's apple groves and making hard cider with a mélange of carefully chosen fruit for many years. The beguiling new flavors I discovered in Terry's ciders (not to mention his long-term loan of *The Apples of New York*, which otherwise would have set me back $400), launched me on the path to apple enlightenment.

Along that path, several pomologists generously shared both their fruit and their knowledge, introducing me to far more apples than I'd ever have met otherwise. This group includes Zeke Goodband at Scott Farm, who has been keeping me in apples for years via Hunger Mountain Coop; Steve Wood at Poverty Lane Orchards, who has been keeping me inebriated for years; Craig and Sharon Campbell of Harmony Orchards; Brad Koehler of Windfall Orchard; the Shelton Family of Vintage Virginia Apples; Todd Parlo of Walden Heights Nursery; and Alan Leonard of Cummins Nursery. John Bunker and Tom Burford helped me see the eternal in the apple tree, and Ben Watson helped connect me with the apple world, as did Lee Kane and Richard Thorpe at Whole Foods. A special shout-out to Thomas Chao and the USDA's Plant Genetic Resources Unit in Geneva, New York. For two glorious days, dawn to dusk, Thomas let me wander the most astonishing apple collection in the world, picking and sampling, a prelapsarian dream. The PGRU's collection is a national treasure of the first order.

I had the treat of working with several talented photographers on this book, including Craig Line in Vermont and Andrea Hubbell in Virginia. A great deal of the personality of this book is thanks to Clare Barboza, whose knack for capturing the essence of each individual apple is quite astounding. Working with natural light and minimal infrastructure, Clare produced masterworks of color and texture that celebrate the everyday richness of the world. She and the apple were a perfect match. The photos were shot in the Kent Tavern and at the homes of Allyson Evans and Greg Labarthe and Gretchen Saries, using pieces from their respective collections. Additional photos were shot Chez Jacobsen, using pieces from the Mary Elder Collection, as well as additional cool stuff lent by East Barre Antique Mall and Julianna Jennings of Antiques at 110 Main. Primary outdoor photography occurred at Harmony Orchard in Tieton, Washington; Windfall Orchard in Cornwall, Vermont; Scott Farm in Dummerston, Vermont; Poverty Lane Orchard in Lebanon, New Hampshire; and Vintage Virginia Apples outside of Charlottesville, Virginia.

Clare's recipe photographs benefited from the styling of Julie Hopper, who played Henry Higgins to my awkwardly brown recipes' Eliza Doolittle. The best two recipes in this book are not mine at all; they were generously contributed by Kate McDermott and Marialisa Calta.

Of course, none of the text and photos would impress without Kathy Belden's editorial eye, Patti Ratchford's wise calls, and Lisa Yee's spot-on design. A talented team, and so easy to work with.

Several editors helped enable my apple obsession by assigning apple stories that furthered my learning. Thank you Kiera Butler at *Mother Jones* (snippets of which piece made it into my introduction), Hunter Lewis at *Southern Living*, Mel Allen at *Yankee*, Justin Paul at *Virtuoso Life*, and Grant Balfour at *Porthole*. Thanks to Catherine Buni for the Cézanne heads-up, and Abigail Carroll for the translation.

A few admirable folks are helping to inspire the current revival of apple culture, and I benefited from their work. Kudos to Sara Grady, Kathleen Frith, and Sabine Hrechdakian at Glynwood in the Hudson Valley; Gary Nabhan and Ben Watson through their work with Slow Food's Ark of Taste; Lee Calhoun in North Carolina; David Buchanan in Portland; John Bunker's whole team at Fedco Trees; Amy Traverso and her fantastic *Apple Lover's Cookbook*; and Diane Flynt and the Apple Corps Project in Virginia. This leaves out the hordes of apple growers and cidermakers whose work has made all of our enthusiasm possible. There has never been a better moment to be an apple geek in America.

APPLE INDEX

RECIPE INDEX

A NOTE ON THE AUTHOR

Rowan Jacobsen is the James Beard Award–winning author of *A Geography of Oysters*, *Fruitless Fall*, *American Terroir*, and other books. He writes for *Harper's*, *Outside*, *Orion*, *Mother Jones*, *Yankee*, and others, and his work has been anthologized in the *Best American Science and Nature Writing* and *Best Food Writing* collections. He lives in Vermont with a wife, a kid, a dog, and a scruffy set of apple trees.

BY THE SAME AUTHOR

Shadows on the Gulf

American Terroir

The Living Shore

Fruitless Fall

A Geography of Oysters

OCT 2014